2018 SQA Specimen and Past Papers with Answers

National 5
GEOGRAPHY

2017 & 2018 Exams
and 2017 Specimen Question Paper

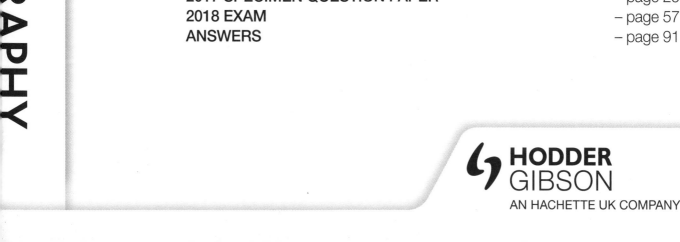

HODDER
GIBSON
AN HACHETTE UK COMPANY

This book contains the official SQA 2017 and 2018 Exams, and the 2017 Specimen Question Paper for National 5 Geography, with associated SQA-approved answers modified from the official marking instructions that accompany the paper.

In addition the book contains study skills advice. This advice has been specially commissioned by Hodder Gibson, and has been written by experienced senior teachers and examiners in line with the new National 5 syllabus and assessment outlines. This is not SQA material but has been devised to provide further guidance for National 5 examinations.

Hodder Gibson is grateful to the copyright holders for permission to use their material. Every effort has been made to trace the copyright holders and to obtain their permission for the use of copyright material. Hodder Gibson will be happy to receive information allowing us to rectify any error or omission in future editions.

Hachette UK's policy is to use papers that are natural, renewable and recyclable products and made from wood grown in sustainable forests. The logging and manufacturing processes are expected to conform to the environmental regulations of the country of origin.

Orders: please contact Bookpoint Ltd, 130 Park Drive, Milton Park, Abingdon, Oxon OX14 4SE. Telephone: (44) 01235 827827. Fax: (44) 01235 400454. Lines are open 9.00–5.00, Monday to Saturday, with a 24-hour message answering service. Visit our website at www.hoddereducation.co.uk. Hodder Gibson can also be contacted directly at hoddergibson@hodder.co.uk

This collection first published in 2018 by
Hodder Gibson, an imprint of Hodder Education,
An Hachette UK Company
211 St Vincent Street
Glasgow G2 5QY

Typeset by Aptara, Inc.

Printed in the UK

A catalogue record for this title is available from the British Library

ISBN: 978-1-5104-5497-2

2 1

2019 2018

Introduction

National 5 Geography

This book of SQA past papers contains the question papers used in the 2017 and 2018 exams (with the answers at the back of the book). A specimen question paper reflecting the content and duration of the revised exam in 2018 is also included. All of the question papers included in the book provide excellent representative practice for the final exams.

Using these papers as part of your revision will help you to develop the vital skills and techniques needed for the exam, and will help you to identify any knowledge gaps you may have.

It is always a very good idea to refer to SQA's website for the most up-to-date course specification documents. These are available at www.sqa.org.uk/sqa/47446

The exam

The course assessment will consist of two parts: a question paper (80 marks) and an assignment (20 marks). The question paper is therefore worth four-fifths of the overall marks of the course assessment, and the assignment one-fifth. The assignment is completed throughout the year and submitted to SQA to be marked in April. Your assignment marks are then added to the marks achieved in the exam paper to give you a final award.

The question paper

The purpose of the question paper is to allow you to demonstrate the skills you have acquired and to reveal the knowledge and understanding you have gained from the topics studied throughout the course. The question paper will give you the chance to show your ability in describing, explaining, matching and evaluating a broad range of geographical information as well as using a variety of maps and demonstrating proficiency in Ordnance Survey (OS) skills. Candidates will complete this question paper in 2 hours and 20 minutes. Questions will be asked on a local, regional and global scale. The question paper has three sections.

Section 1: Physical Environments

This section is worth 30 marks. Candidates will answer a mixture of limited/extended-response questions by using the knowledge, understanding and skills learned throughout the course. In this section there is a choice. Candidates should answer **either** Question 1: glaciation/coasts or Question 2: rivers/limestone. This will be dependent on the subjects taught at your school. Some topics you could be asked to answer questions on include **Weather, Landscape formations** within Scotland and/or the UK, and **Land use management** – conflicts and solutions. In this

section you may also be examined on your Ordnance Survey skills using a map.

Section 2: Human Environments

This section is worth 30 marks. As in Section 1, candidates will answer a mixture of limited/extended-response questions by using the knowledge, understanding and skills learned throughout the course. Candidates should answer **all** questions in this section, which will be drawn from both the developed and developing world. Some topics you could be asked questions on include **Population** (development indicators, population distribution, factors affecting birth rates and death rates), **Urban** (land use characteristics in cities in the developed world, recent developments in developed world cities, strategies to improve shanty towns) and **Rural** (changes in rural landscapes in both the developed and developing world). In this section you may also be examined on your Ordnance Survey skills using a map.

Section 3: Global Issues

This section is worth 20 marks, made up of two 10-mark questions. Again, candidates will answer these questions by using the knowledge, understanding and skills learned throughout the course. In this section there is a choice of questions. Candidates should answer **two** questions from a choice of six. Your choice will be dependent on the topics taught at your school. The choice of topics is: **Climate change**, **Natural regions**, **Environmental hazards**, **Trade and globalisation**, **Tourism** and **Health**.

Types of questions

The main types of questions used in the paper are: **Describe, Explain, Give reasons, Match, Give advantages and/or disadvantages,** and **Give map evidence.**

Describe questions

You must make a number of relevant, factual points. These should be key points taken from a given source, for example a map, diagram or table.

Explain or Give reasons questions

You should make a number of points giving clear reasons for a given situation. The command word "explain" will be used when you are asked to demonstrate knowledge and understanding. Sometimes the command words "give reasons" may be used as an alternative to "explain".

Match questions

You are asked to match two sets of variables, for example to match features to a correct grid reference.

Advantages and/or disadvantages questions

You should select relevant advantages or disadvantages of a proposed development, for example the location of a new shopping centre, and demonstrate your understanding of the significance of the proposal.

Give map evidence questions

You should look for evidence on the map and make clear statements to support your answer.

Some tips for revising

- To be best prepared for the examination, organise your notes into sections. Try to work out a schedule for studying with a programme which includes the sections of the syllabus you intend to study.
- Organise your notes into checklists and revision cards.
- Make sure you have a copy of the examination timetable and have planned a schedule for studying.
- Try to avoid leaving your studying to a day or two before the exam. Also try to avoid cramming your studies into the night before the examination, and especially avoid staying up late to study.
- One useful technique when revising is to use summary note cards on individual topics.
- Make use of past paper questions to test your knowledge and skills. Go over your answers and give yourself a mark for every correct point you make when comparing your answer with your notes.
- If you work with a classmate, try to mark each other's practice answers.
- Practise your diagram-drawing skills and your writing skills. Ensure that your answers are clearly worded. Try to develop the points that you make in your answers.

Some tips for the exam

- Do not write lists, even if you are running out of time. You will lose marks. If the question asks for an opinion based on a choice, for example on the suitability of a particular site or area for a development, do not be afraid to refer to negative points such as why the alternatives are not as good. You will get credit for this.
- Arrive at the examination in plenty of time with the appropriate equipment – pen, pencil, rubber and ruler.
- Carefully read the instructions on the paper and at the beginning of each part of the question.
- Answer all of the compulsory questions in each paper you sit.
- Use the number of marks as a guide to the length of your answer.
- Try to include examples in your answer wherever possible. If asked for diagrams, draw clear, labelled diagrams.
- Read the question instructions very carefully. If the question asks you to "describe", make sure that this is what you do.

- If you are asked to "explain", you must use phrases such as "due to", "this happens because" and "this is a result of". If you describe, rather than explain, you will lose most of the marks for that question.
- If you finish early, do not leave the exam. Use the remaining time to check your answers and go over any questions which you have partially answered, especially Ordnance Survey map questions.
- Practise drawing diagrams which may be included in your answers, for example corries or pyramidal peaks.
- Make sure that you have read the instructions on the question carefully and that you have avoided needless errors. For example, answering the wrong sections or failing to explain when asked to, or perhaps forgetting to refer to a named area or case study.
- One technique which you might find helpful when answering 5- or 6-mark questions is to "brainstorm" possible points for your answer. You can write these down in a list at the start of your answer. As you go through your answer, you can double-check with your list to ensure that you have put as much into your answer as you can. This stops you from coming out of the exam and being annoyed that you forgot to mention an important point.

Common errors

Markers of the external examination often remark on errors which occur frequently in candidates' answers. These include the following:

Lack of sufficient detail

- Many candidates fail to provide sufficient detail in answers, often by omitting reference to specific examples, or not elaborating or developing points made in their answer. As noted above, a good guide to the amount of detail required is the number of marks given for the question. If, for example, the total marks offered is 6, then you should make at least six valid points.

Listing

- If you write a simple list of points rather than fuller statements in your answer, you will automatically lose marks. For example, in a 4-mark question, you will obtain only 1 mark for a list.
- The same rule applies to a simple list of bullet points. However, if you couple bullet points with some detailed explanation, you could achieve full marks.

Irrelevant answers

- You must read the question instructions carefully so as to avoid giving answers which are irrelevant to the question. For example, if you are asked to "explain" and you simply "describe", you will lose marks. If you are asked for a named example and you do not provide one, you will forfeit marks.

Repetition

- You should be careful not to repeat points already made in your answer. These will not gain any further marks. You may feel that you have written a long answer, but it may contain the same basic information repeated again and again. Unfortunately, these repeated statements will be ignored by the marker.

Good luck!

Remember that the rewards for passing National 5 Geography are well worth it! Your pass will help you to get the future you want for yourself. In the exam, be confident in your own ability. If you're not sure how to answer a question, trust your instincts and just give it a go anyway – keep calm and don't panic! GOOD LUCK!

Study Skills – what you need to know to pass exams!

General exam revision: 20 top tips

When preparing for exams, it is easy to feel unsure of where to start or how to revise. This guide to general exam revision provides a good starting place, and, as these are very general tips, they can be applied to all your exams.

1. Start revising in good time.

Don't leave revision until the last minute – this will make you panic and it will be difficult to learn. Make a revision timetable that counts down the weeks to go.

2. Work to a study plan.

Set up sessions of work spread through the weeks ahead. Make sure each session has a focus and a clear purpose. What will you study, when and why? Be realistic about what you can achieve in each session, and don't be afraid to adjust your plans as needed.

3. Make sure you know exactly when your exams are.

Get your exam dates from the SQA website and use the timetable builder tool to create your own exam schedule. You will also get a personalised timetable from your school, but this might not be until close to the exam period.

4. Make sure that you know the topics that make up each course.

Studying is easier if material is in manageable chunks – why not use the SQA topic headings or create your own from your class notes? Ask your teacher for help on this if you are not sure.

5. Break the chunks up into even smaller bits.

The small chunks should be easier to cope with. Remember that they fit together to make larger ideas. Even the process of chunking down will help!

6. Ask yourself these key questions for each course:

- Are all topics compulsory or are there choices?
- Which topics seem to come up time and time again?
- Which topics are your strongest and which are your weakest?

Use your answers to these questions to work out how much time you will need to spend revising each topic.

7. Make sure you know what to expect in the exam.

The subject-specific introduction to this book will help with this. Make sure you can answer these questions:

- How is the paper structured?
- How much time is there for each part of the exam?
- What types of question are involved? These will vary depending on the subject so read the subject-specific section carefully.

8. Past papers are a vital revision tool!

Use past papers to support your revision wherever possible. This book contains the answers and mark schemes too – refer to these carefully when checking your work. Using the mark scheme is useful; even if you don't manage to get all the marks available first time when you first practise, it helps you identify how to extend and develop your answers to get more marks next time – and of course, in the real exam.

9. Use study methods that work well for you.

People study and learn in different ways. Reading and looking at diagrams suits some students. Others prefer to listen and hear material – what about reading out loud or getting a friend or family member to do this for you? You could also record and play back material.

10. There are three tried and tested ways to make material stick in your long-term memory:

- Practising – e.g. rehearsal, repeating
- Organising – e.g. making drawings, lists, diagrams, tables, memory aids
- Elaborating – e.g. incorporating the material into a story or an imagined journey

11. Learn actively.

Most people prefer to learn actively – for example, making notes, highlighting, redrawing and redrafting, making up memory aids, or writing past paper answers. A good way to stay engaged and inspired is to mix and match these methods – find the combination that best suits you. This is likely to vary depending on the topic or subject.

12. Be an expert.

Be sure to have a few areas in which you feel you are an expert. This often works because at least some of them will come up, which can boost confidence.

13. Try some visual methods.

Use symbols, diagrams, charts, flashcards, post-it notes etc. Don't forget – the brain takes in chunked images more easily than loads of text.

14. Remember – practice makes perfect.

Work on difficult areas again and again. Look and read – then test yourself. You cannot do this too much.

15. Try past papers against the clock.

Practise writing answers in a set time. This is a good habit from the start but is especially important when you get closer to exam time.

16. Collaborate with friends.

Test each other and talk about the material – this can really help. Two brains are better than one! It is amazing how talking about a problem can help you solve it.

17. Know your weaknesses.

Ask your teacher for help to identify what you don't know. Try to do this as early as possible. If you are having trouble, it is probably with a difficult topic, so your teacher will already be aware of this – most students will find it tough.

18. Have your materials organised and ready.

Know what is needed for each exam:

- Do you need a calculator or a ruler?
- Should you have pencils as well as pens?
- Will you need water or paper tissues?

19. Make full use of school resources.

Find out what support is on offer:

- Are there study classes available?
- When is the library open?
- When is the best time to ask for extra help?
- Can you borrow textbooks, study guides, past papers, etc.?
- Is school open for Easter revision?

20. Keep fit and healthy!

Try to stick to a routine as much as possible, including with sleep. If you are tired, sluggish or dehydrated, it is difficult to see how concentration is even possible. Combine study with relaxation, drink plenty of water, eat sensibly, and get fresh air and exercise – all these things will help more than you could imagine. Good luck!

NATIONAL 5

2017

FRIDAY, 26 MAY

1:00 PM – 2:45 PM

Total marks — 60

SECTION 1 — PHYSICAL ENVIRONMENTS — 20 marks

Attempt EITHER Question 1 OR Question 2. ALSO attempt Questions 3 and 4.

SECTION 2 — HUMAN ENVIRONMENTS — 20 marks

Attempt Questions 5, 6, 7 and 8.

SECTION 3 — GLOBAL ISSUES — 20 marks

Attempt any TWO of the following.

Question   9 — Climate Change

Question 10 — Impact of Human Activity on the Natural Environment

Question 11 — Environmental Hazards

Question 12 — Trade and Globalisation

Question 13 — Tourism

Question 14 — Health

Credit will be given for appropriately labelled sketch maps and diagrams.

Write your answers clearly in the answer booklet provided. In the answer booklet you must clearly identify the question number you are attempting.

Use blue or black ink.

Before leaving the examination room you must give your answer booklet to the Invigilator; if you do not, you may lose all the marks for this paper.

MARKS

SECTION 1 — PHYSICAL ENVIRONMENTS — 20 marks
Attempt EITHER Question 1 OR Question 2
ALSO attempt Questions 3 and 4

Question 1: Glaciated Upland

Diagram Q1: Glacial contour patterns

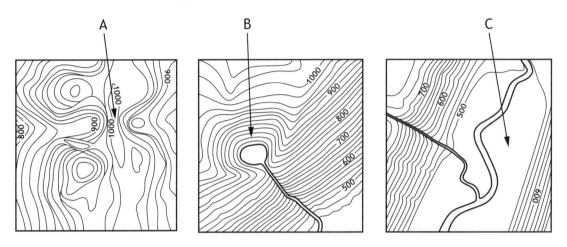

(a) Match the letters on Diagram Q1 with the correct glacial features below.
     Choose from:

U-shaped valley	corrie	pyramidal peak	arête	
				3

(b) **Explain** the processes involved in the formation of a U-shaped valley.
     You may use a diagram(s) in your answer.   4

(c) **Explain** different ways in which people use glaciated landscapes.   4

NOW ATTEMPT QUESTIONS 3 AND 4

MARKS

DO NOT ATTEMPT THIS QUESTION IF YOU HAVE ALREADY ANSWERED QUESTION 1

Question 2: Upland Limestone

Diagram Q2: Upland limestone landscape

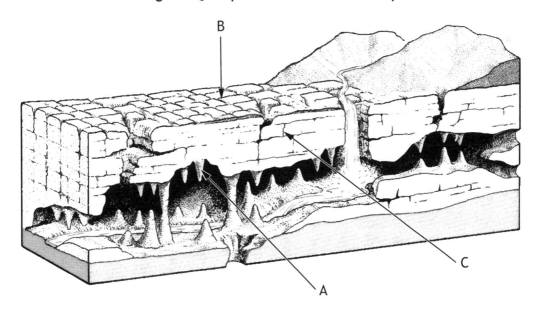

(a) Match the letters on Diagram Q2 with the correct limestone features below.

Choose from:

stalactite	stalagmite	clint	grike	joint	bedding plane	3

(b) **Explain** the formation of a limestone pavement.
You may use a diagram(s) in your answer.   4

(c) **Explain** different ways in which people use limestone landscapes.   4

NOW ATTEMPT QUESTIONS 3 AND 4

[Turn over

MARKS

Question 3

Diagram Q3: Synoptic chart for 12.00 on 28th December 2014

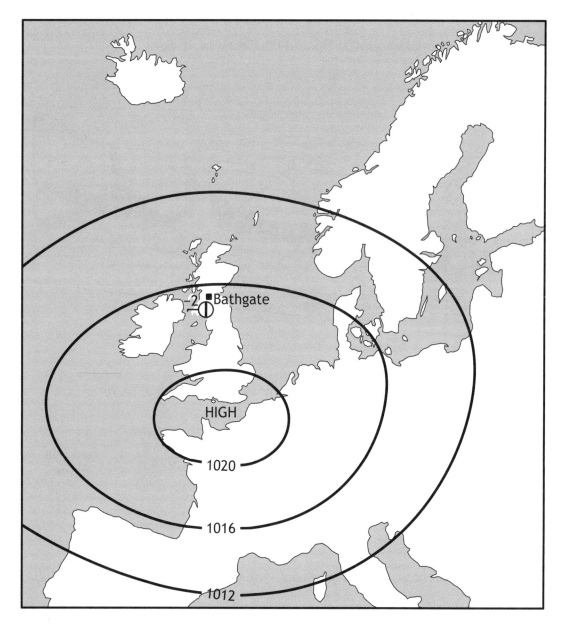

Study Diagram Q3.

Give reasons for the weather conditions at Bathgate on 28th December 2014.   5

Question 4

Anticyclones bring different weather conditions throughout the year.

Describe the benefits **and** problems of an anticyclone in **summer**.   4

MARKS

SECTION 2 — HUMAN ENVIRONMENTS — 20 marks
Attempt Questions 5, 6, 7 and 8

Question 5

Study the Ordnance Survey map extract (Item A) of the Edinburgh area.

Match the grid references with the correct urban land use zone.

Grid references: **2568, 2573, 2671**

Choose from the urban land use zones below.

- **CBD**
- **new industry**
- **new housing**
- **old housing**

3

Question 6

Study the Ordnance Survey map extract (Item A) of the Edinburgh area.

There is a plan to build new housing in grid square 2667.

Using map evidence, **explain** why this area is suitable for new housing.

5

[Turn over

Question 7

Diagram Q7: World Population Density

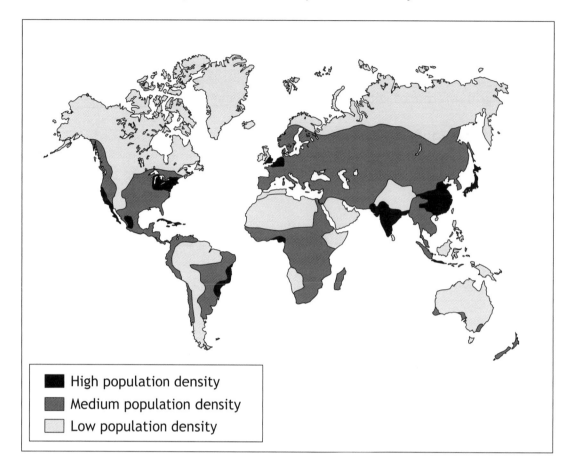

High population density
Medium population density
Low population density

Look at Diagram Q7.

Explain why there are areas of different population density across the world.

Your answer should refer to both physical **and** human factors.

6

Question 8

Diagram Q8: Shanty Town

De Visu / Shutterstock.com

Look at Diagram Q8.

Referring to an area you have studied, **describe** different ways shanty towns are being improved.

6

[Turn over

SECTION 3 — GLOBAL ISSUES — 20 marks
Attempt any TWO questions

MARKS

Question 9 — Climate Change

Diagram Q9A: Average Global Temperature Change 1996−2016

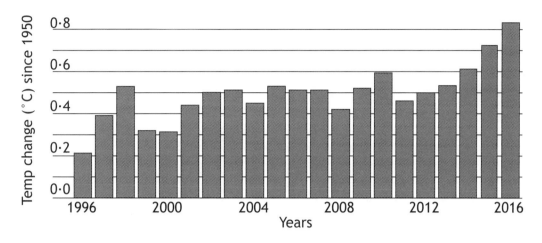

(a) Study Diagram Q9A.

Describe, in detail, average global temperature change from 1996 to 2016.   4

Diagram Q9B: Newspaper Headline

(b) Look at Diagram Q9B.

Explain, in detail, strategies used to minimise future climate change.   6

[Turn over

MARKS

Question 10 — Impact of Human Activity on the Natural Environment

Diagram Q10

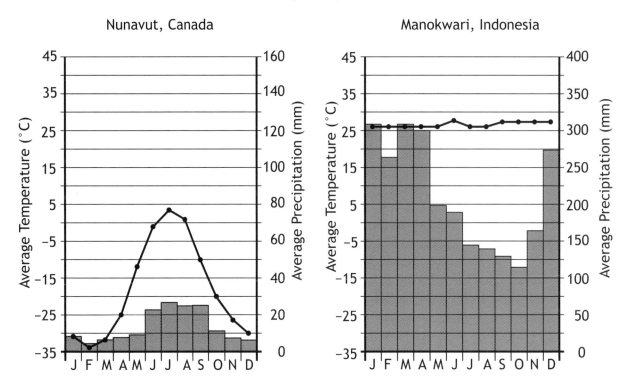

(a) Study Diagram Q10.

Use the information in Diagram Q10 to **describe**, **in detail**, the differences between the two climates shown.

4

(b) For a named **tundra** or **equatorial** area which you have studied, **explain** the impact of human activity on people **and** the environment.

6

MARKS

Question 11 — Environmental Hazards

Diagram Q11A: World Distribution of Tropical Storms

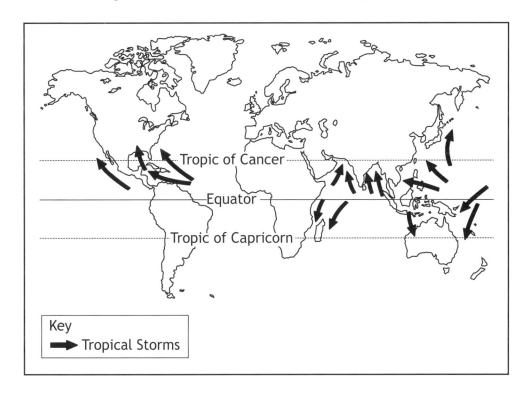

(a) Study Diagram Q11A.

Describe, in detail, the distribution of tropical storms.

4

Diagram Q11B: A Tropical Storm hits coastal town

Fabio Lamanna / Shutterstock.com

(b) Look at Diagram Q11B.

For a tropical storm you have studied, **explain in detail** the impacts of the storm on people **and** the environment.

6

MARKS

Question 12 — Trade and Globalisation

Diagram Q12A: Percentage Share of World Trade 1995–2010

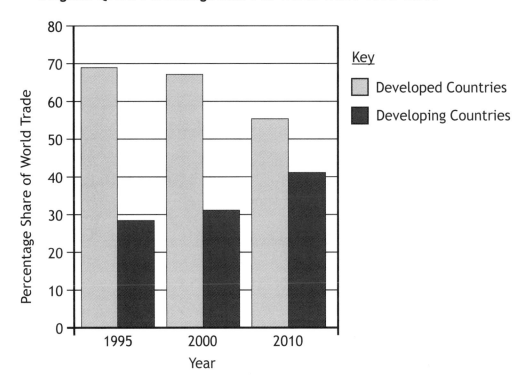

(a) Study Diagram Q12A.

Describe, **in detail**, the changes in percentage share of world trade from 1995–2010.

4

Diagram Q12B: Trade Report

Trade Report
Trade Inequality increases between Developing and Developed countries in 2015.

(b) Look at Diagram Q12B.

Explain, **in detail**, the causes of inequalities in trade between developed and developing countries.

6

MARKS

Question 13 — Tourism

Diagram Q13A: Origin of Tourists Visiting Scotland (thousands)

COUNTRY	2006	2010	2014
USA	475	275	418
Germany	278	253	343
France	229	196	190
Australia	133	147	158
Netherlands	114	135	149
Canada	161	98	122
Ireland	224	185	113
Spain	142	139	101
Rest of World	976	913	1,106
TOTAL	2,732	2,341	2,700

(a) Study Diagram Q13A.

Describe, in detail, the changes in the number of tourists visiting Scotland from different countries between 2006 and 2014.

4

Diagram Q13B: Quote from a tour operator

> "Mass tourism has increased since the 1950s with many locations at home and abroad experiencing a record number of visitors year on year."

(b) Look at Diagram Q13B.

Give reasons for the increase in mass tourism.

6

[Turn over

Question 14 — Health

Diagram Q14A: Percentage Change in Death Rates from Malaria, 2000–2013

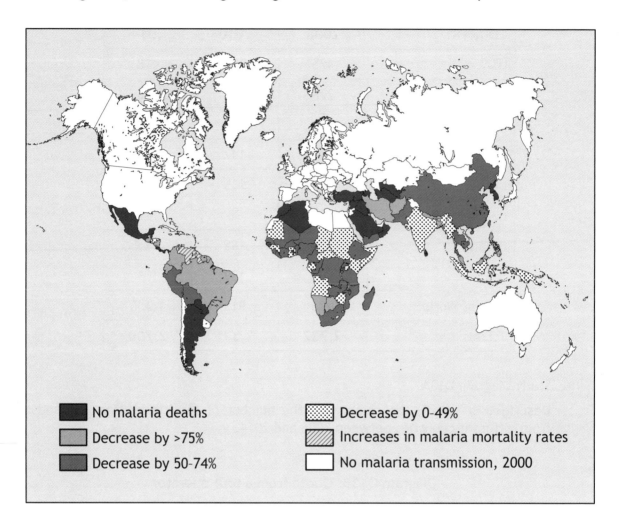

No malaria deaths

Decrease by >75%

Decrease by 50-74%

Decrease by 0-49%

Increases in malaria mortality rates

No malaria transmission, 2000

(a) Study Diagram Q14A.

   Describe, in detail, the changes in death rates from malaria.   4

MARKS

Question 14 — Health (continued)

Diagram Q14B: Selected Developing World Diseases

cholera
pneumonia
malaria
kwashiorkor

(b) Look at Diagram Q14B.

Choose **one** disease from Diagram Q14B above.

For the disease you have chosen, explain the impact on people **and** the countries affected.

6

[END OF QUESTION PAPER]

MARKS

[BLANK PAGE]

DO NOT WRITE ON THIS PAGE

National Qualifications 2017

X733/75/21

Geography
Ordnance Survey Map
Item A

FRIDAY, 26 MAY

1:00 PM — 2:45 PM

The colours used in the printing of these map extracts are indicated in the four little boxes at the top of the map extract. Each box should contain a colour; if any does not, the map is incomplete and should be returned to the Invigilator.

Ordnance Survey

1:50 000 Scale
Landranger Series

ROADS AND PATHS

Not necessarily rights of way

Junction number
Service area — Elevated
M1 (S)
Unfenced

Motorway (dual carriageway)

A 470

Primary Route
(A network of recommended through routes
which complement the motorway system)

Dual carriageway

A 493

Main road

Footbridge

Road under construction

Secondary road

B 4518

Narrow road with passing places

A 855 Bridge B 885

Road generally more than 4m wide

Road generally less than 4m wide

Path / Other road, drive or track

Gradient: steeper than 20% (1 in 5),
14% to 20% (1 in 7 to 1 in 5)

Gates, Road tunnel

Ferry P Ferry V

Ferry (passenger), Ferry (vehicle)

RAILWAYS

Track multiple or single
Track under construction
Siding
Tunnel, cuttings
Narrow gauge, tramway
or light rail system

Bridges, footbridge
Level crossing LC
Viaduct, embankment
Station, (a) principal
Light rail station

WATER FEATURES

Marsh or salting
Towpath Lock
Aqueduct Canal
Weir Ford
Footbridge Bridge
Lake
Canal (dry)

Cliff
Slopes
Shingle
Flat rock
Sand Lighthouse
Dunes (disused)
Beacon Lighthouse (in use)
Normal tidal limit
Low water mark
Mud
High water mark

HEIGHTS

1 metre = 3·2808 feet

30

Contours are at 10 metres
vertical interval

·144

Heights are to the nearest
metre above mean sea level

Where two heights are shown, the first is
the height of the natural ground in the
location of the triangulation pillar, and the
second (in brackets) to a separate point
which is the natural summit.

ROCK FEATURES

Outcrop
Cliff
Scree

PUBLIC RIGHTS OF WAY

············ Footpath
– – – – – Bridleway
– · – · – · Restricted byway (not for
use by mechanically
propelled vehicles)
–+–+–+– Byway open to all traffic

The symbols show the defined route so far
as the scale of mapping will allow.

The representation on this map of any other
road, track or path is no evidence of the
existence of a right of way. Not shown on
maps of Scotland

Danger Area

Firing and Test Ranges
in the area. Danger!
Observe warning notices.

OTHER PUBLIC ACCESS

· · · · Other route with public access
(not normally shown in urban
areas). Alignments are based on
the best information available.
These routes are not shown on
maps of Scotland.

On-road cycle route
Traffic-free cycle route
4 National Cycle Network number
8 Regional Cycle Network number
National Trail, Scotland's Great Trails,
European Long Distance Path and
selected Recreational Routes

BOUNDARIES

–·–·–·– National
–··–··–·· District
– ·· – ·· County, Unitary Authority,
Metropolitan District
or London Borough

National Park

ANTIQUITIES

+ Site of antiquity
⚔ Site of Battle (with date)
☆ ···· Visible earthwork
VILLA Roman
Castle Non-Roman

TOURIST INFORMATION

Camp site / caravan site
Garden/aboretum
Golf course or links
i Information centre (all year / seasonal)
Nature reserve
P Parking, Park and ride (all year / seasonal)
Picnic site
Recreation / leisure / sports centre
Selected places of tourist interest
Phone, public / emergency
Viewpoint
V Visitor centre
Walks / Trails
World Heritage site or area
▲ Youth hostel

LAND FEATURES

Electricity transmission line
(pylons shown at standard spacing)
Pipe line
(arrow indicates direction of flow)
ruin Buildings
Important building (selected)
Bus or coach station
Current or with tower
former place
of worship with spire, minaret or dome
+ Place of worship
Glass structure
H Heliport
Triangulation pillar
Mast
Wind pump
Wind turbine
Windmill with or without sails
Graticule intersection at 5' intervals
Cutting, embankment
Landfill site or slag/spoil heap
Coniferous wood
Non-coniferous wood
Mixed wood
Orchard
Park or ornamental ground
Forestry Commission land
National Trust (always open / limited access,
observe local signs)
Natural Resources Wales
National Trust for Scotland (always open /
limited access, observe local signs)

ABBREVIATIONS

Br Bridge MS Milestone
Cemy Cemetery Mus Museum
CG Cattle grid P Post office
CH Clubhouse PC Public convenience (in rural areas)
Fm Farm PH Public house
Hospl Hospital Sch School
Ho House TH Town Hall, Guildhall or equivalent
MP Milepost Univ University

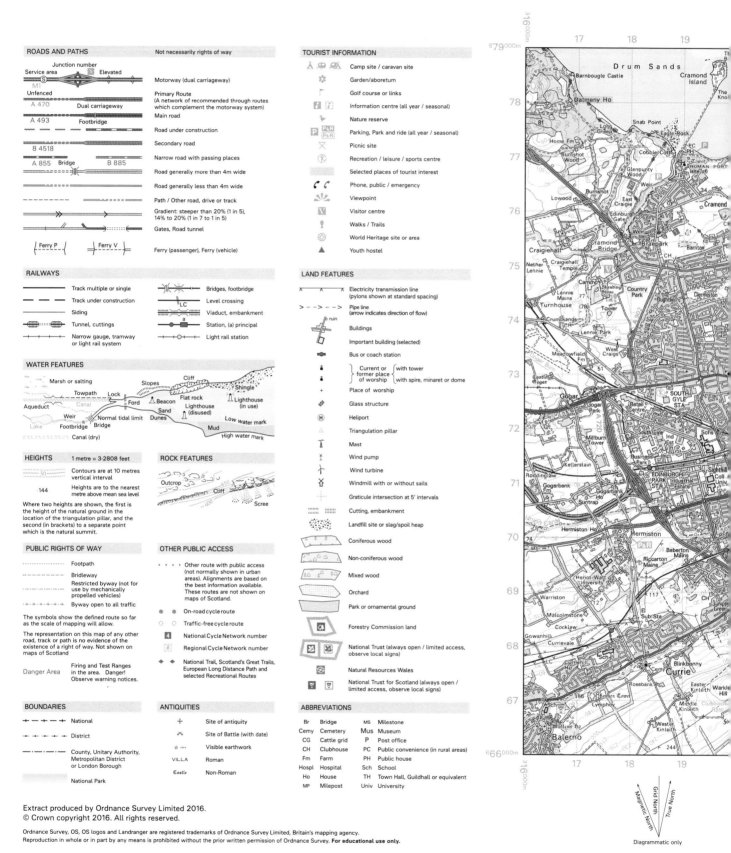

Four colours should appear above; if not then please return to the invigilator.

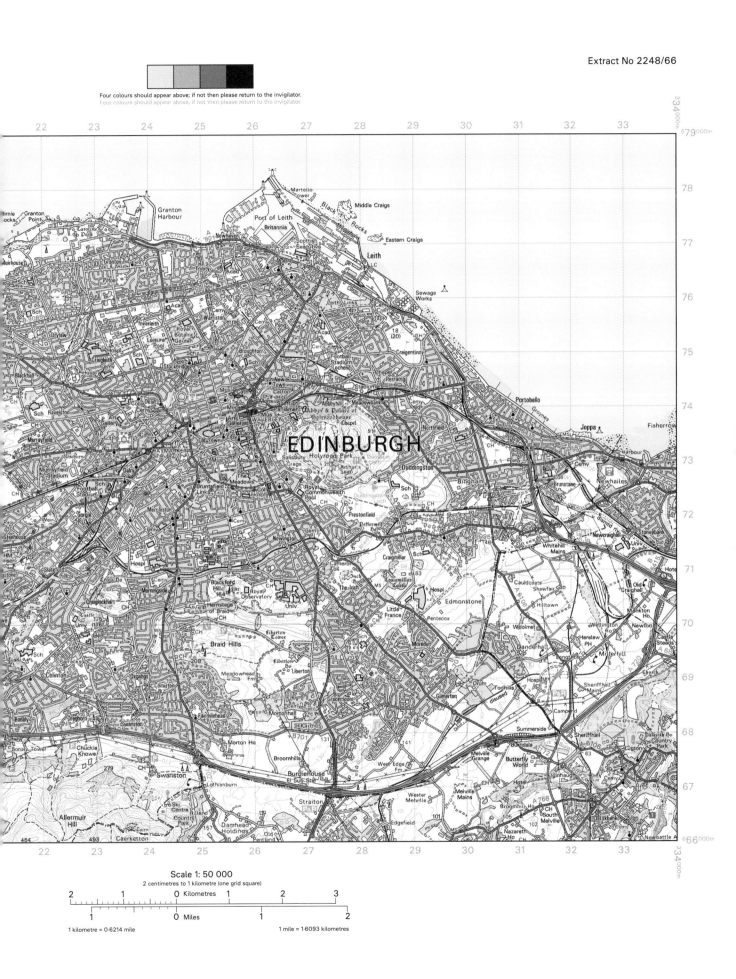

Scale 1: 50 000

2 centimetres to 1 kilometre (one grid square)

1 kilometre = 0·6214 mile   1 mile = 1·6093 kilometres

[BLANK PAGE]

DO NOT WRITE ON THIS PAGE

NATIONAL 5

2017 Specimen
Question Paper

National Qualifications
SPECIMEN ONLY

S833/75/11

Geography

Date — Not applicable

Duration — 2 hours 20 minutes

Total marks — 80

SECTION 1 — PHYSICAL ENVIRONMENTS — 30 marks

Attempt **EITHER** question 1 **OR** question 2

THEN attempt questions 3 to 6.

SECTION 2 — HUMAN ENVIRONMENTS — 30 marks

Attempt ALL questions.

SECTION 3 — GLOBAL ISSUES — 20 marks

Attempt any **TWO** of the following.

Question 11 — Climate change

Question 12 — Natural regions

Question 13 — Environmental hazards

Question 14 — Trade and globalisation

Question 15 — Tourism

Question 16 — Health

You will receive credit for appropriately labelled sketch maps and diagrams.

Write your answers clearly in the answer booklet provided. In the answer booklet you must clearly identify the question number you are attempting.

Use **blue** or **black** ink.

Before leaving the examination room you must give your answer booklet to the Invigilator; if you do not, you may lose all the marks for this paper.

MARKS

SECTION 1 — PHYSICAL ENVIRONMENTS — 30 marks
Attempt EITHER question 1 OR question 2
THEN questions 3 to 6

Question 1 — Glaciated landscapes

(a) Study the Ordnance Survey map extract (Item A) of the Brecon Beacons area.

Using grid references, **describe** the evidence shown on the map which suggests that this is an area of **upland glaciated scenery**.   **4**

(b) **Explain** the formation of a **U-shaped valley**.

You may use a diagram(s) in your answer.   **4**

Now attempt questions 3 to 6

MARKS

Do not attempt question 2 if you have already answered question 1

Question 2 — Rivers and valleys

(a) Study the Ordnance Survey map extract (Item A) of the Brecon Beacons area.

Describe the physical features of the Afon (River) Nedd Fechan **and** its valley between 905175 and 900092. You should use grid references in your answer.   **4**

(b) **Explain** the formation of an **ox-bow lake**.

You may use a diagram(s) in your answer.   **4**

Now attempt questions 3 to 6

Question 3

Item Q3: Quote from a local landowner

"This area has the potential for a variety of different land uses, including farming, forestry, recreation/tourism, water storage/supply, industry and renewable energy."

Study Item Q3 and the Ordnance Survey map extract (Item A) of the Brecon Beacons area.

Choose **two** different land uses mentioned in Item Q3.

Using map evidence, **explain** how the area shown on the map extract is suitable for your chosen land uses.

5

MARKS

Question 4

Diagram Q4: Synoptic chart, 0800 hours, 10th March

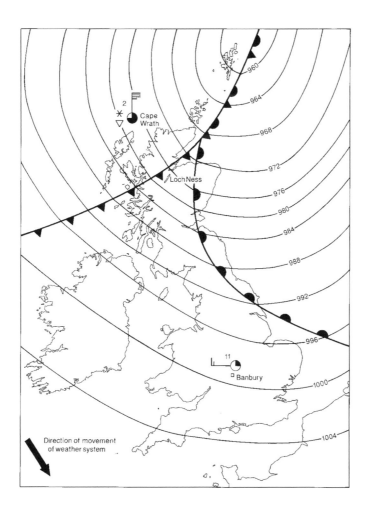

Study Diagram Q4.

(a) **Describe**, in detail, the differences in the weather between Cape Wrath and Banbury at 0800 hours on 10th March.   4

(b) At 0800 hours on 10th March a group of secondary school students are about to set off on a walk into the mountains near Loch Ness. After seeing the weather chart in Diagram Q4, they decide to cancel their walk at the last minute.

Why might conditions have been unsuitable for their expedition? Give reasons.   5

[Turn over

MARKS

Question 5

Diagram Q5: Air masses affecting the British Isles

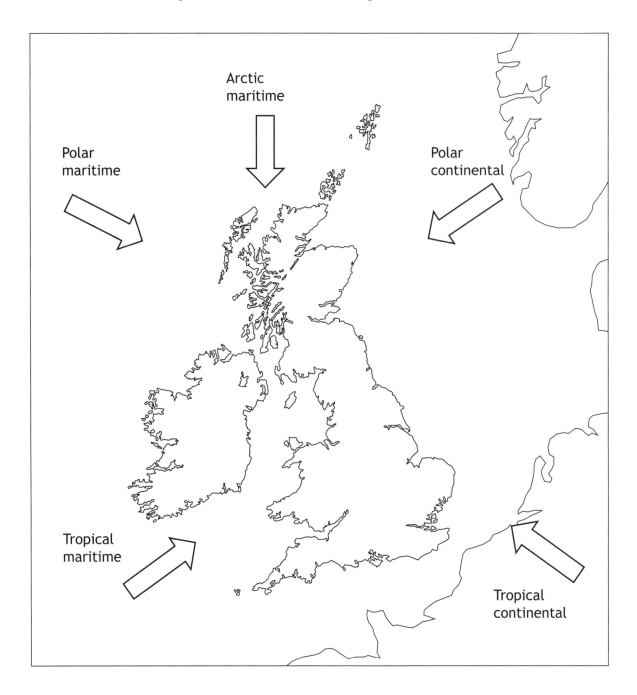

Look at Diagram Q5.

Describe how a long period with a **tropical continental** air mass **in summer** would affect the people of the British Isles.

3

MARKS

Question 6

Diagram Q6: Selected land uses

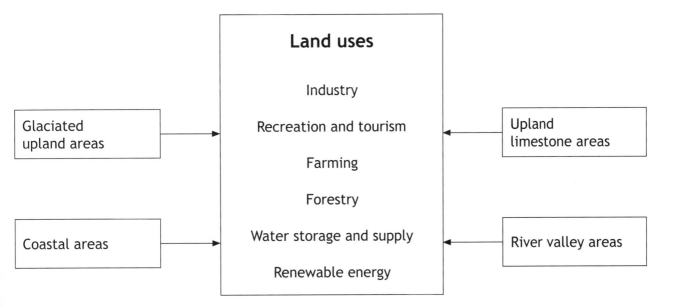

Look at Diagram Q6.

(i) For a named area you have studied, **explain**, **in detail**, ways in which **two** different land uses may be in conflict with each other.

(ii) **Suggest** possible solutions to these conflicts.

5

[Turn over

MARKS

SECTION 2 — HUMAN ENVIRONMENTS — 30 marks
Attempt ALL questions

Question 7

Diagram Q7

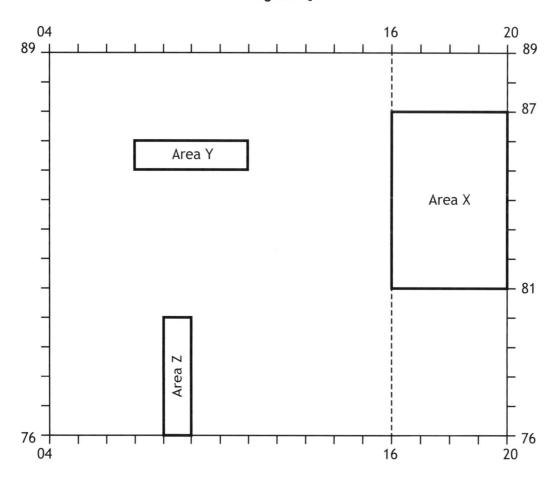

Study the Ordnance Survey map extract (Item B) of the Birmingham area and Diagram Q7 above.

(a) Give map evidence to show that part of the Central Business District (CBD) of Birmingham is found in grid square 0786.   3

(b) Find Area X on Diagram Q7 and the map extract (Item B).

Birmingham airport, a golf course, a business park and a housing area are found in Area X on the rural/urban fringe of Birmingham. Using map evidence **explain** why such developments are found there.   5

(c) The Russell family have three young children and are buying a house in Birmingham. They have narrowed down their search to two areas of the city — Area Y (Balsall Heath and Sparkbrook) or Area Z (Highter's Heath and Drake's Cross).

Which area, Y or Z, should they choose? Using **detailed map evidence, give reasons** to support your chosen area.   6

MARKS

Question 8

Diagram Q8: Developments in farming

GM crops

Biofuel

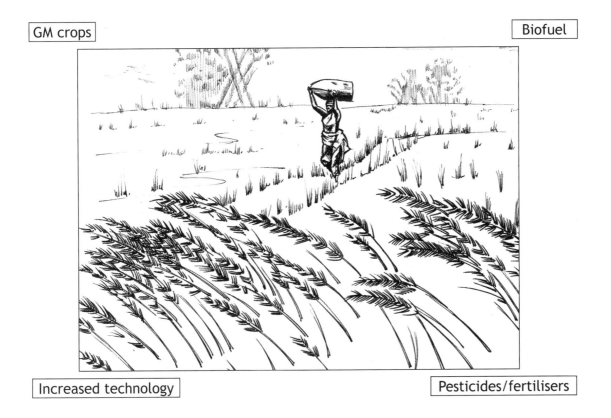

Increased technology

Pesticides/fertilisers

Look at Diagram Q8.

Explain how recent developments in agriculture in developing countries are helping farmers.

4

[Turn over

Question 9   MARKS

Diagram Q9: Demographic transition model

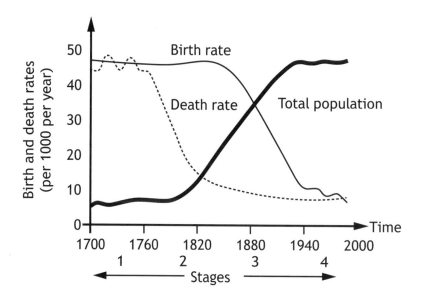

Look at Diagram Q9.

Countries such as the United Kingdom have experienced significant population change, as shown in Diagram Q9.

Explain why this population change has happened. You must refer to factors affecting birth and death rates in stages 2, 3 and 4 of Diagram Q9.   6

Question 10

Table Q10: Selected development indicators

Country	Life expectancy (yrs)	Access to safe drinking water (%)	Literacy rate (%)	% of workforce employed in agriculture
Bolivia	69	90	96	32
Chad	50	51	40	80
Finland	81	100	100	4
Mali	56	77	39	80
Netherlands	81	100	100	2
Uganda	55	79	78	40

Study Table Q10.

Choose **two** of the development indicators shown.

For the **two** that you have chosen, **explain**, **in detail**, why they are useful in helping to show a country's level of development.   6

MARKS

SECTION 3 — GLOBAL ISSUES — 20 marks
Attempt any TWO questions

MARKS

Question 11: Climate change

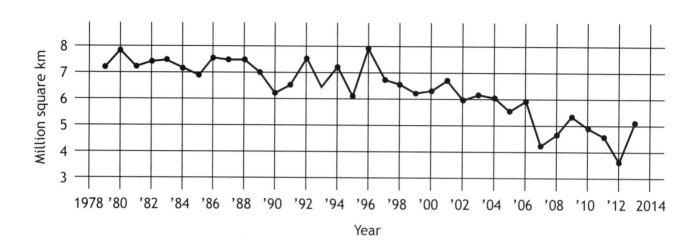

Diagram Q11: Area of Arctic Sea ice (1979–2013)

Study Diagram Q11.

(a) **Describe, in detail**, the changes in the area of Arctic Sea ice.   4

(b) Melting sea ice is one effect of climate change.

    Explain some other effects of climate change.   6

MARKS

Question 12: Natural regions

Diagram Q12: Recent deforestation rates worldwide

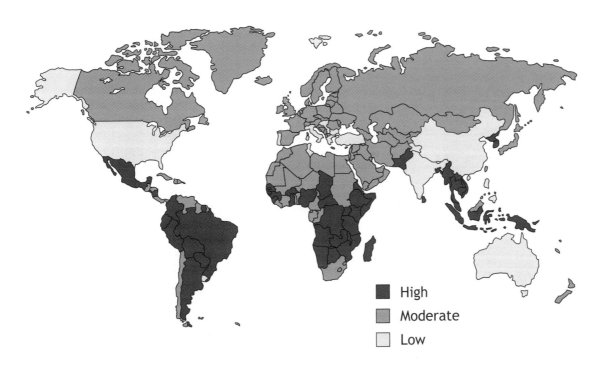

(a) Study Diagram Q12.

Describe, **in detail**, deforestation rates worldwide.　　4

(b) **Explain** the management strategies which can be used to minimise the impact of human activity in the tundra.　　6

[Turn over

Question 13: Environmental hazards

Diagram Q13A: Number of volcanic eruptions per decade 1910–2010

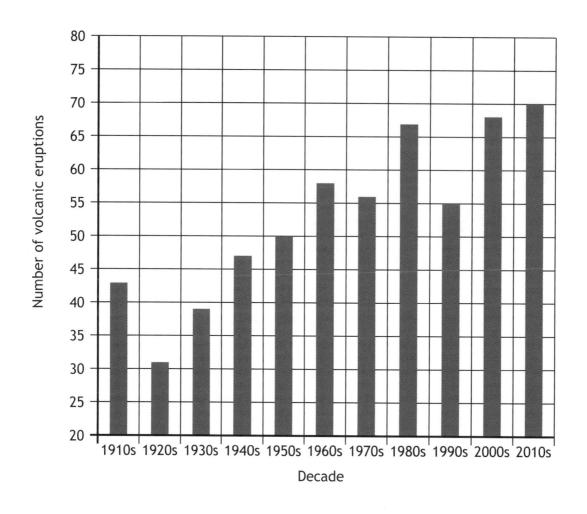

(a) Study Diagram Q13A.

Describe, **in detail**, the changes in the number of volcanic eruptions between 1910–2010.

4

MARKS

Question 13 (continued)

Item Q13B: Pico de Fogo volcano, Cape Verde

After nearly 20 years of inactivity, the Pico de Fogo awakened with a violent eruption on the 23rd of November 2014.

(b) Look at Item Q13B.

For a volcanic eruption you have studied, **explain**, **in detail**, the impacts of the eruption on people and the landscape.

6

[Turn over

MARKS

Question 14: Trade and globalisation

Diagram Q14A: World exports by region

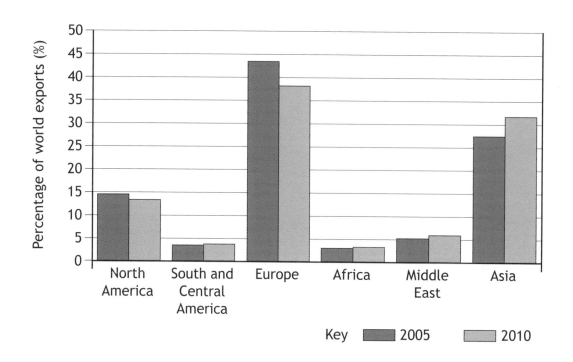

Key ▬ 2005   ▭ 2010

(a) Study Diagram Q14A.

   Describe, in detail, the changes in world exports from 2005 to 2010.   4

Item Q14B: Collecting Fairtrade coffee beans

(b) Look at Item Q14B.

   Explain how buying Fairtrade products helps people in the developing world.   6

MARKS

Question 15: Tourism

Diagram Q15A: Top ten world tourist destinations (millions of visitors per year)

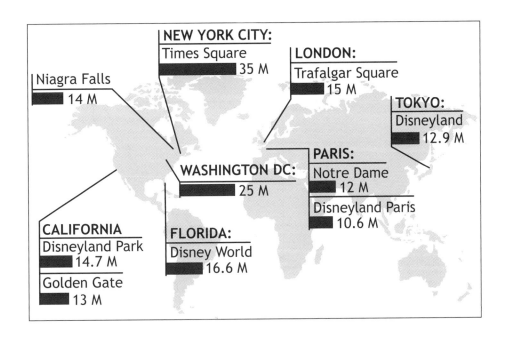

(a) Study Diagram Q15A.

Describe, in detail, the distribution of the top ten world tourist destinations.　4

Item Q15B — Mass tourism on an Italian beach

(b) Look at Item Q15B.

Describe the effects of mass tourism on people and the environment.　6

MARKS

Diagram Q16: Ebola cases in selected African countries April–Oct 2014

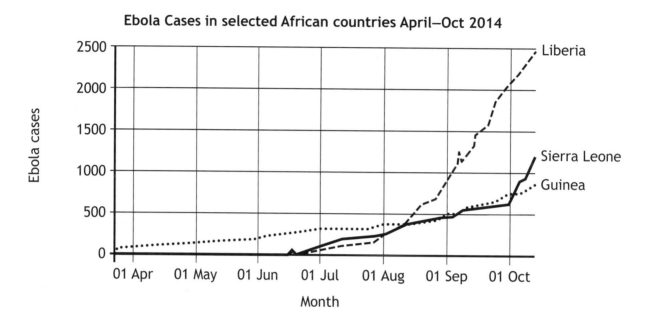

(a) Study Diagram Q16.

 Describe, in detail, the changes in Ebola cases in the **three** named African countries.   4

(b) **Explain** the causes of either heart disease **or** cancer **or** asthma.   6

[END OF SPECIMEN QUESTION PAPER]

National Qualifications
SPECIMEN ONLY

S833/75/11

Geography
Ordnance Survey Map
Item A

Date — Not applicable

Duration — 2 hours 20 minutes

The colours used in the printing of these map extracts are indicated in the four little boxes at the top of the map extract. Each box should contain a colour; if any does not, the map is incomplete and should be returned to the Invigilator.

Extract No 2213/160

3

Four colours should appear above; if not then please return to the invigilator.
Four colours should appear above; if not then please return to the invigilator.

ROADS AND PATHS
Not necessarily rights of way

Junction number Service area Elevated	
S M1 Unfenced	Motorway (dual carriageway)
A 470 Dual carriageway	Primary Route (recommended through route)
A 493	Main road
Footbridge	Road under construction
B 4518	Secondary road
A 855 Bridge B 885	Narrow road with passing places
	Road generally more than 4m wide
	Road generally less than 4m wide
	Path / Other road, drive or track
	Gradient: steeper than 20% (1 in 5), 14% to 20% (1 in 7 to 1 in 5)
	Gates, Road tunnel
Ferry P Ferry V	Ferry (passenger), Ferry (vehicle)

RAILWAYS

	Track multiple or single		Bridges, footbridge
	Track under construction	LC	Level crossing
	Siding		Viaduct, embankment
	Tunnel, cuttings		Station, (a) principal
	Light rapid transit system, narrow gauge or tramway		Light rapid transit system station

WATER FEATURES

Marsh or salting
Slopes Cliff
Towpath Lock
Shingle
Flat rock
Lighthouse (in use)
Aqueduct Canal Beacon
Ford
Weir Normal tidal limit
Lighthouse (disused)
Lake Footbridge Sand Dunes
Low water mark
Canal (dry) Bridge
Mud
High water mark

HEIGHTS
1 metre = 3·2808 feet

30   Contours are at 10 metres vertical interval

·144   Heights are to the nearest metre above mean sea level

Where two heights are shown the first height is to the base of the triangulation pillar and the second (in brackets) to the highest natural point of the hill

ROCK FEATURES

Outcrop
Cliff
Scree

PUBLIC RIGHTS OF WAY

....	Footpath
– – – – –	Bridleway
–·–·–·–	Restricted byway
·+·+·+·	Byway open to all traffic

The symbols show the defined route so far as the scale of mapping will allow.

The representation on this map of any other road, track or path is no evidence of the existence of a right of way. Not shown on maps of Scotland

Danger Area   Firing and Test Ranges in the area. Danger! Observe warning notices.

OTHER PUBLIC ACCESS

• • • •   Other route with public access (not normally shown in urban areas). Alignments are based on the best information available. These routes are not shown on maps of Scotland.

● ●   On-road cycle route

○ ○   Traffic-free cycle route

4   National Cycle Network number

8   Regional Cycle Network number

◆ ◆   National Trail, European Long Distance Path, Long Distance Route, selected Recreational Routes

BOUNDARIES

–+–·–+–	National
–+–+–+–	District
–·–·–·–	County, Unitary Authority, Metropolitan District or London Borough
	National Park

ANTIQUITIES

+	Site of antiquity
⚔	Battlefield (with date)
☆	Visible earthwork
VILLA	Roman
Castle	Non-Roman

TOURIST INFORMATION

⚐ ⚑	Camp site / caravan site
✿	Garden
	Golf course or links
i i	Information centre (all year / seasonal)
	Nature reserve
P P&R P&R	Parking, Park and ride (all year / seasonal)
✗	Picnic site
	Recreation / leisure / sports centre
	Selected places of tourist interest
☏ ☏	Telephone, public / roadside assistance
	Viewpoint
V	Visitor centre
!	Walks / Trails
◎	World Heritage site or area
▲	Youth hostel

LAND FEATURES

×——×——×	Electricity transmission line (pylons shown at standard spacing)
>–·–>–·–>	Pipe line (arrow indicates direction of flow)
in ruin	Buildings
	Important building (selected)
	Bus or coach station
	Current or former place of worship { with tower / with spire, minaret or dome }
+	Place of worship
⌀	Glass structure
H	Heliport
△	Triangulation pillar
I	Mast
Y	Wind pump, wind turbine
✕	Windmill with or without sails
+	Graticule intersection at 5' intervals
	Cutting, embankment
	Landfill site or slag/spoil heap
	Coniferous wood
	Non-coniferous wood
	Mixed wood
	Orchard
	Park or ornamental ground
	Forestry Commission land
	National Trust (always open / limited access, observe local signs)
	National Trust for Scotland (always open / limited access, observe local signs)

ABBREVIATIONS

Br	Bridge	MS	Milestone
Cemy	Cemetery	Mus	Museum
CG	Cattle grid	P	Post office
CH	Clubhouse	PC	Public convenience (in rural areas)
Fm	Farm	PH	Public house
Ho	House	Sch	School
MP	Milepost	TH	Town Hall, Guildhall or equivalent

1:50 000 Scale
Landranger Series

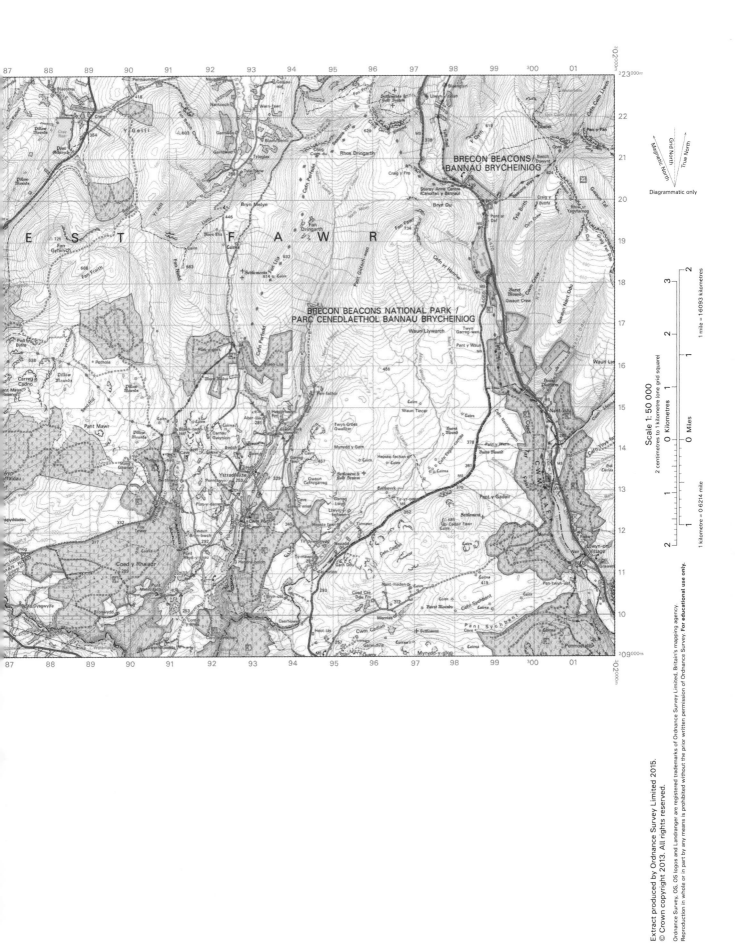

Scale 1: 50 000

2 centimetres to 1 kilometre (one grid square)

1 kilometre = 0·6214 mile

1 mile = 1·6093 kilometres

[BLANK PAGE]

DO NOT WRITE ON THIS PAGE

National Qualifications
SPECIMEN ONLY

S833/75/11

**Geography
Ordnance Survey Map
Item B**

Date — Not applicable

Duration — 2 hours 20 minutes

The colours used in the printing of these map extracts are indicated in the four little boxes at the top of the map extract. Each box should contain a colour; if any does not, the map is incomplete and should be returned to the Invigilator.

Extract No 2142/139

1:50 000 Scale
Landranger Series

ROADS AND PATHS

Not necessarily rights of way

Service area — Junction number — Elevated
M1
Motorway (dual carriageway)

Unfenced
A 470
Primary Route (recommended through route)

A 493 — Dual carriageway
Main road

Footbridge
Road under construction

B 4518
Secondary road

A 855 — Bridge — B 885
Narrow road with passing places

Road generally more than 4m wide

Road generally less than 4m wide

Path / Other road, drive or track

Gradient: steeper than 20% (1 in 5),
14% to 20% (1 in 7 to 1 in 5)

Gates, Road tunnel

Ferry P — Ferry V
Ferry (passenger), Ferry (vehicle)

RAILWAYS

Track multiple or single — Bridges, footbridge

Track under construction — Level crossing
LC

Siding — Viaduct, embankment

Tunnel, cuttings — Station, (a) principal
a

Light rapid transit system, narrow gauge or tramway — Light rapid transit system station

WATER FEATURES

Marsh or salting — Slopes — Cliff
Towpath — Lock — Shingle
Aqueduct — Canal — Ford — Flat rock — Lighthouse (in use)
Weir — Beacon — Lighthouse (disused)
Lake — Footbridge — Bridge — Sand — Dunes — Normal tidal limit — Low water mark
Mud — High water mark
Canal (dry)

HEIGHTS

1 metre ≈ 3·2808 feet

Contours are at 10 metres
50 vertical interval

·144 Heights are to the nearest
metre above mean sea level

Where two heights are shown the first height is to
the base of the triangulation pillar and the second
(in brackets) to the highest natural point of the hill

ROCK FEATURES

Outcrop
Cliff
Scree

PUBLIC RIGHTS OF WAY

· · · · · · · · · · Footpath

— — — — — Bridleway

– · – · – · – · Restricted byway

-+-+-+-+- Byway open to all traffic

The symbols show the defined route so far as the
scale of mapping will allow.

The representation on this map of any other road,
track or path is no evidence of the existence of a
right of way. Not shown on maps of Scotland

Danger Area — Firing and Test Ranges in the
area. Danger! Observe
warning notices.

OTHER PUBLIC ACCESS

· · · · Other route with public access
(not normally shown in urban areas).
Alignments are based on the best
information available. These routes
are not shown on maps of Scotland.

⊕ On-road cycle route

○ Traffic-free cycle route

4 National Cycle Network number

8 Regional Cycle Network number

◆ National Trail, European Long Distance
Path, Long Distance Route, selected
Recreational Routes

BOUNDARIES

—+—+—+ National

—+—+—+ District

—·—·—·— County, Unitary Authority,
Metropolitan District
or London Borough

National Park

ANTIQUITIES

+ Site of antiquity

⚔ Battlefield (with date)

☆ ···· Visible earthwork

VILLA Roman

Castle Non-Roman

TOURIST INFORMATION

Camp site / caravan site

Garden

Golf course or links

i i Information centre (all year / seasonal)

Nature reserve

P P&R Parking, Park and ride (all year / seasonal)
P&R

⤬ Picnic site

Recreation / leisure / sports centre

Selected places of tourist interest

Telephone, public / roadside assistance

Viewpoint

V Visitor centre

Walks / Trails

World Heritage site or area

▲ Youth hostel

LAND FEATURES

Electricity transmission line
(pylons shown at standard spacing)

> – > – > Pipe line
(arrow indicates direction of flow)

ruin Buildings

Important building (selected)

Bus or coach station

Current or } with tower
former place
of worship } with spire, minaret or dome

+ Place of worship

Glass structure

(H) Heliport

Triangulation pillar

Mast

Wind pump, wind turbine

Windmill with or without sails

Graticule intersection at 5' intervals

Cutting, embankment

Landfill site or slag/spoil heap

Coniferous wood

Non-coniferous wood

Mixed wood

Orchard

Park or ornamental ground

Forestry Commission land

National Trust (always open / limited access,
observe local signs)

National Trust for Scotland (always open /
limited access, observe local signs)

ABBREVIATIONS

Br	Bridge	MS	Milestone
Cemy	Cemetery	Mus	Museum
CG	Cattle grid	P	Post office
CH	Clubhouse	PC	Public convenience (in rural areas)
Fm	Farm	PH	Public house
Ho	House	Sch	School
MP	Milepost	TH	Town Hall, Guildhall or equivalent

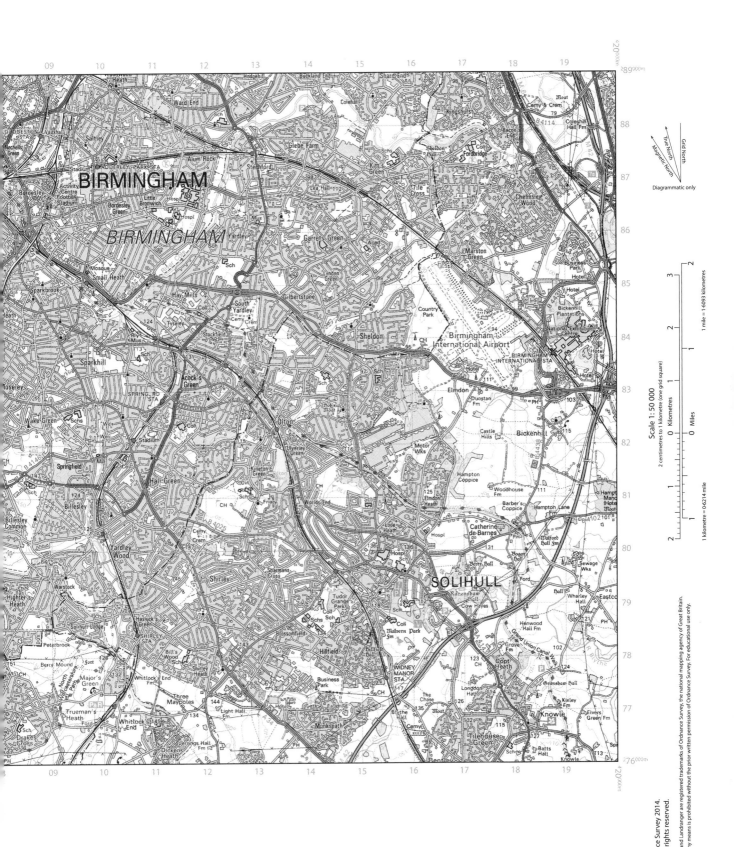

Scale 1 : 50 000

2 centimetres to 1 kilometre (one grid square)

[BLANK PAGE]

DO NOT WRITE ON THIS PAGE

NATIONAL 5

2018

National Qualifications 2018

X833/75/11

Geography

TUESDAY, 1 MAY

1:00 PM – 3:20 PM

Total marks — 80

SECTION 1 — PHYSICAL ENVIRONMENTS — 30 marks

Attempt **EITHER** question 1 **OR** question 2.

THEN attempt questions 3 to 7.

SECTION 2 — HUMAN ENVIRONMENTS — 30 marks

Attempt ALL questions.

SECTION 3 — GLOBAL ISSUES — 20 marks

Attempt any **TWO** of the following.

Question 13 — Climate change

Question 14 — Natural regions

Question 15 — Environmental hazards

Question 16 — Trade and globalisation

Question 17 — Tourism

Question 18 — Health

You will receive credit for appropriately labelled sketch maps and diagrams.

Write your answers clearly in the answer booklet provided. In the answer booklet you must clearly identify the question number you are attempting.

Use **blue** or **black** ink.

Before leaving the examination room you must give your answer booklet to the Invigilator; if you do not, you may lose all the marks for this paper.

MARKS

SECTION 1 — PHYSICAL ENVIRONMENTS — 30 marks
Attempt EITHER question 1 OR question 2
THEN questions 3 to 7

Question 1 — Coastal landscapes

(a) Study the Ordnance Survey map extract (Item A) of the Strathy area.

Match these grid references with the correct coastal features.

Grid references: **827694, 812681 and 843662.**

Choose from features: **cliff; stack; sand spit; arch.**   3

(b) Explain the formation of a sand spit. You may use a diagram(s) in your answer.   4

Now attempt questions 3 to 7

MARKS

Do not attempt question 2 if you have already answered question 1

Question 2 — Rivers and their valleys

    (a) Study the Ordnance Survey map extract (Item A) of the Strathy area.

          Match these grid references with the correct river features.

          Grid references: **893618, 883627 and 895589.**

          Choose from features: **v-shaped valley; flood plain; meander; ox-bow lake.**   3

    (b) Explain the formation of a meander. You may use a diagram(s) in your answer.   4

Now attempt questions 3 to 7

[Turn over

Question 3

Diagram Q3A: Cross-section GR 858624 to GR 910590

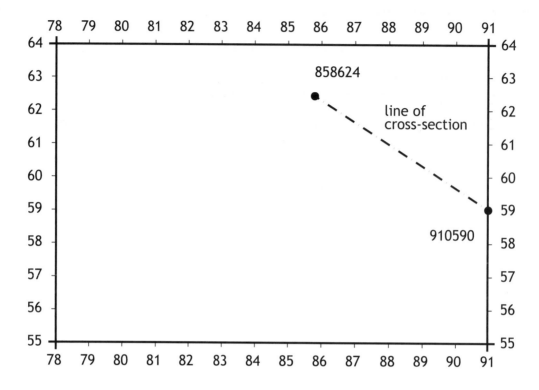

Question 3 (continued)

Diagram Q3B: Cross-section

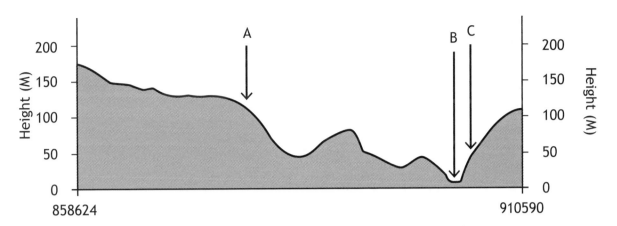

Study the Ordnance Survey map extract (Item A) of the Strathy area and Diagrams Q3A and Q3B.

Match the letters A—C with the correct features.   3

Choose from the features below.

> Halladale River; track; forestry; electricity transmission lines.

[Turn over

Question 4

Diagram Q4: Quote from local council official

> "This area has the potential for a variety of different land uses including
> - farming
> - forestry
> - recreation/tourism
> - water storage/supply
> - industry
> - renewable energy."

Study Diagram Q4 and the Ordnance Survey map extract (Item A) of the Strathy area.

Choose **two** different land uses listed in Diagram Q4.

Using map evidence, **explain** how the area shown on the map extract is suitable for your two chosen land uses.

5

MARKS

Question 5

Diagram Q5: Selected land uses

Look at Diagram Q5.

Choose **one** landscape type from Diagram Q5.

For a named area you have studied, **explain in detail** ways in which land use conflicts may be managed.

6

[Turn over

MARKS

Question 6

Diagram Q6: Average annual UK temperatures

⑪ Average annual UK temperatures

Look at Diagram Q6.

Explain the factors which affect average temperatures in the UK.   4

MARKS

Question 7

Diagram Q7: Synoptic chart for Monday, 16 April 2016 at 8am

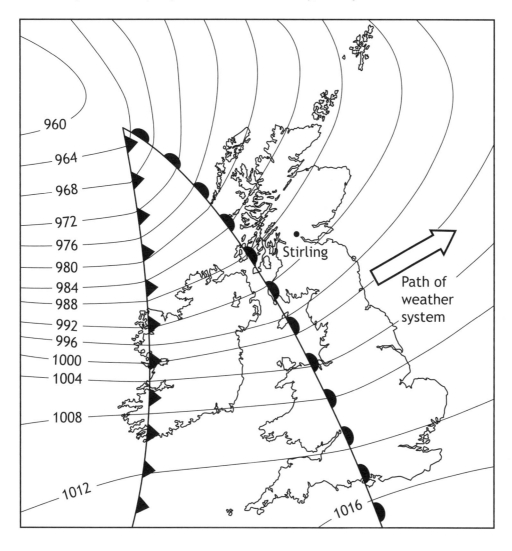

Study Diagram Q7 above.

Explain the changes that will take place in the weather in Stirling over the next 24 hours.

5

[Turn over

MARKS

SECTION 2 — HUMAN ENVIRONMENTS — 30 marks
Attempt ALL questions

Question 8

Study the Ordnance Survey map extract (Item B) of the Oxford area.

Measure the three distances (A, B and C) between the places shown in the table.

Match your answers for A, B and C with the distances given below.

A	From the public telephone in Henwood (4702) to the school near Rose Hill (5303)
B	From Forest Farm (5410) to the church in Stanton St John (5709)
C	From Waterperry Gardens (6206) to the College (5502)

Choose from the following distances:

8·25 km   3·75 km   6·25 km   12·5 km   3

MARKS

Question 9

Diagram Q9

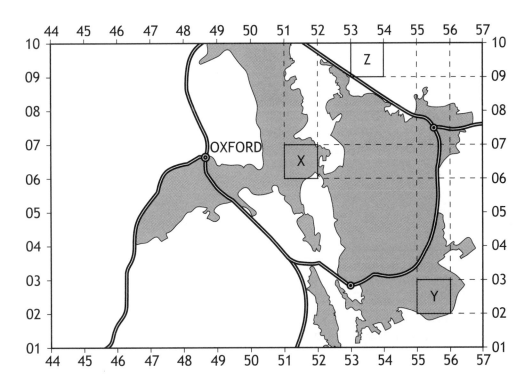

Study Diagram Q9 and the Ordnance Survey map extract (Item B) of the Oxford area.

(a) Give map evidence to explain why Area X is the CBD **and** Area Y is the suburbs.   **4**

(b) There is a proposal to build a new supermarket in grid square 5309 (Area Z).
Give the advantages **and** disadvantages of Area Z for this development.
You must use map evidence in your answer.   **5**

[Turn over

MARKS

Question 10

Diagram Q10: Inner city problems

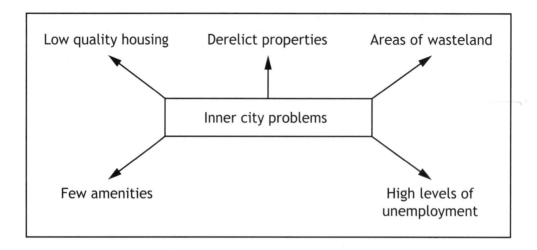

Look at Diagram Q10.

Referring to a **developed** world city you have studied, give reasons for recent changes which have taken place in the inner city.

6

Question 11

Diagram Q11: Changes in farming in developing countries

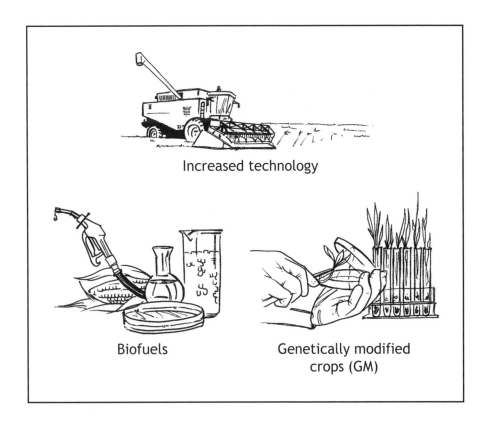

Look at Diagram Q11.

Describe, in detail, the effects of recent changes in farming on people **and** the landscape in **developing** countries.

You must mention at least **two** recent changes.   4

[Turn over

MARKS

Question 12

Diagram Q12A: Gross national income 2015 (total per country)

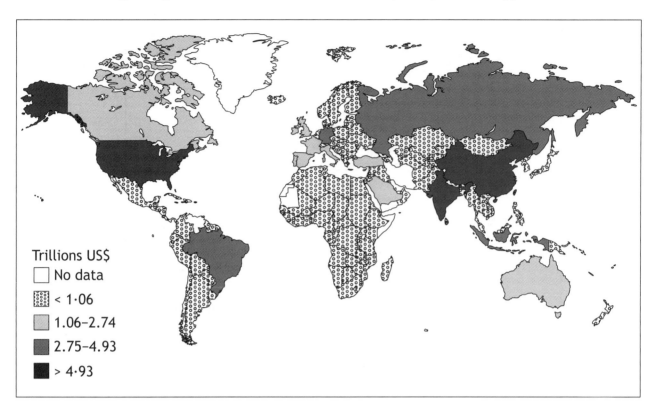

Trillions US$
- No data
- < 1·06
- 1·06–2·74
- 2·75–4·93
- > 4·93

(a) Study Diagram Q12A.

Describe, **in detail,** the different Gross National Incomes in 2015 worldwide. 4

Diagram Q12B: Selected indicators of development

Social indicator	Economic indicator
Number of people per doctor	% of people working in agriculture
% of people who can read and write	Average income per person per year
Number of births per 1,000 women per year	Gross Domestic Product (GDP) per year

(b) Look at Diagram Q12B.

Choose **one** social and **one** economic indicator of development shown in the table.

Explain how your two chosen indicators show the level of development in a country. 4

SECTION 3 — GLOBAL ISSUES — 20 marks
Attempt any TWO questions

[Turn over

Question 13: Climate change

Diagram Q13A: Worldwide greenhouse gas emissions
(1990 to 2010)

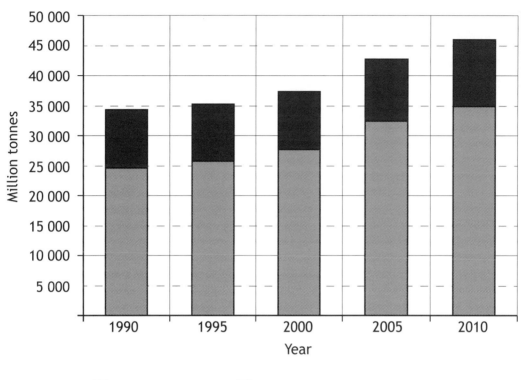

(a) Study Diagram Q13A.

Describe, in detail, the changes in greenhouse gas emissions between 1990 and 2010.   4

MARKS

Question 13 (continued)

Diagram Q13B: Online newspaper report

(b) Look at Diagram Q13B.

Explain the physical **and** human causes of climate change.　　6

[Turn over

MARKS

Question 14: Natural regions

Diagram Q14A: Changes in deforestation and world population: 1900 to 2010

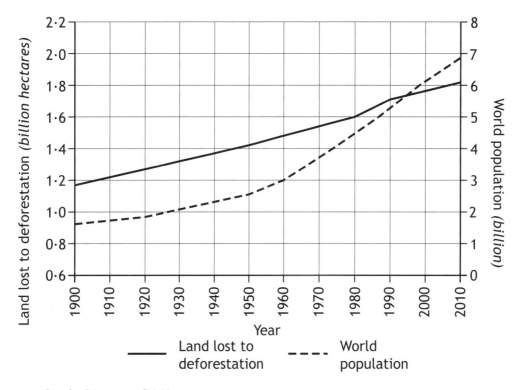

(a) Study Diagram Q14A.

Describe, in detail, changes in deforestation **and** population between 1900 and 2010.

4

MARKS

Question 14 (continued)

Diagram Q14B: Examples of vegetation adaptations in equatorial rainforests

Fan palm

Buttress roots

Vegetation adaptation rainforest

Lianas

Strangler fig

Diagram Q14C: Examples of vegetation adaptations in the tundra

Woolly seed covers

Cup shaped flowers

Vegetation adaptation tundra

Cotton grass

Plants growing close together

(b) Look at Diagrams Q14B and Q14C.

Explain ways in which vegetation has adapted to the environment in **either** the rainforest **or** the tundra.

6

[Turn over

Question 15: Environmental hazards

Diagram Q15A:
Estimated damage (US$ millions) caused by natural disasters 1990 to 2012

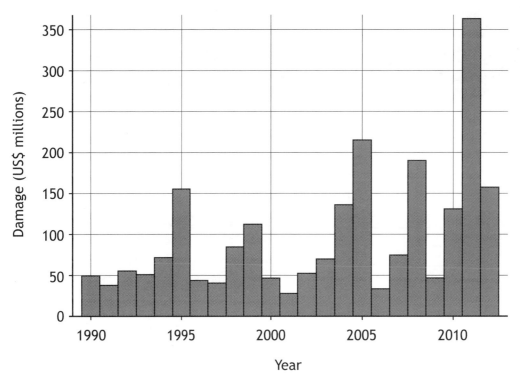

(a) Study Diagram Q15A.

Describe, in detail, the changes in estimated damage caused by natural disasters from 1990 to 2012.

4

MARKS

Question 15 (continued)

Diagram Q15B: Natural hazards in the news

(b) Look at Diagram Q15B.

For the volcano(es) that you have studied, **explain in detail** the strategies used
to prepare for and reduce the effects of an eruption.   **6**

[Turn over

MARKS

Question 16: Trade and globalisation

Diagram Q16: Number of Fair Trade employees 2013 to 2014

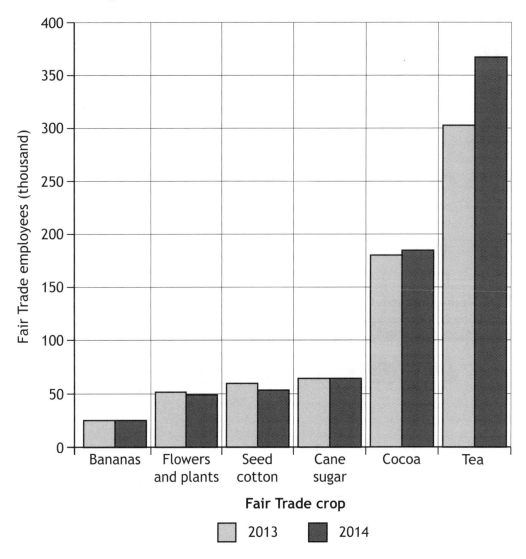

(a) Study Diagram Q16.

Describe, in detail, the changes in the number of Fair Trade employees from 2013 to 2014.   4

(b) Referring to a country or countries you have studied, **explain** how Fair Trade can help people.   6

MARKS

Question 17: Tourism

Diagram Q17: World heritage sites facing threat from tourism

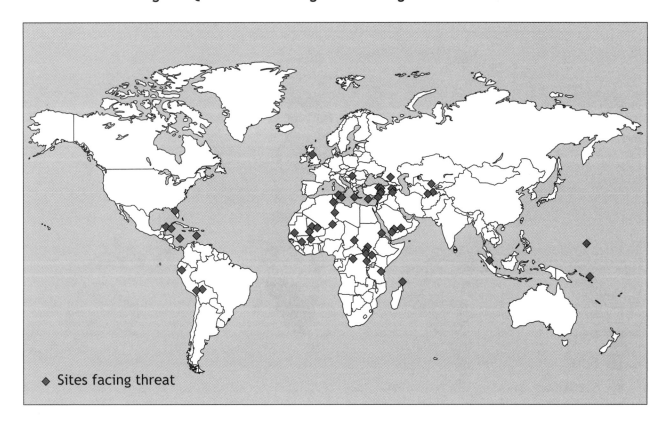

♦ Sites facing threat

(a) Study Diagram Q17.

   Describe, in detail, the location of world heritage sites facing threat from tourism.   **4**

(b) For named areas that you have studied, **describe** ways eco-tourism can be managed.   **6**

[Turn over for next question

MARKS

Question Q18: Health

Diagram Q18: Adult HIV infection rate by country 2013

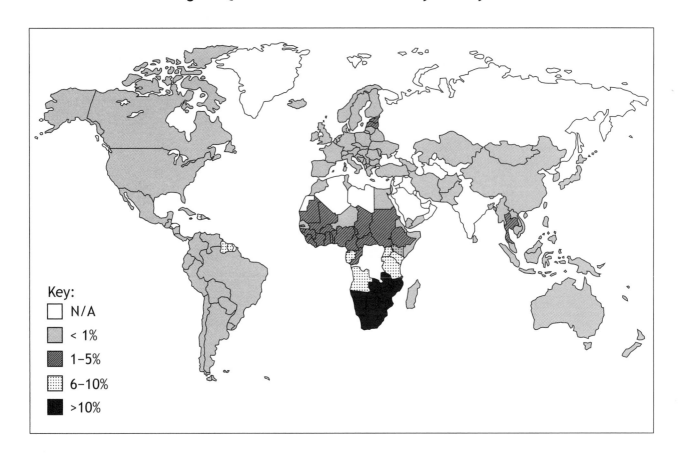

Key:
- □ N/A
- ▨ < 1%
- ▨ 1–5%
- ▨ 6–10%
- ■ >10%

(a) Study Diagram Q18.

Describe, **in detail,** the global distribution of HIV/AIDS infection amongst adults.   4

(b) **Explain** the effects of HIV/AIDs on the populations of **developing** countries.   6

[END OF QUESTION PAPER]

National Qualifications 2018

X833/75/21

Geography
Ordnance Survey Map
Item A

TUESDAY, 1 MAY
1:00 PM — 3:20 PM

The colours used in the printing of these map extracts are indicated in the four little boxes at the top of the map extract. Each box should contain a colour; if any does not, the map is incomplete and should be returned to the Invigilator.

Ordnance Survey

1:50 000 Sc
Landranger

ROADS AND PATHS

Not necessarily rights of way

Junction number
Service area | Elevated
M1

Motorway (dual carriageway)

Unfenced
A 470　Dual carriageway

Primary Route (recommended through route)

A 493　Footbridge

Main road

Road under construction

B 4518

Secondary road

A 855　Bridge　B 885

Narrow road with passing places

Road generally more than 4m wide

Road generally less than 4m wide

Path / Other road, drive or track

Gradient: steeper than 20% (1 in 5),
14% to 20% (1 in 7 to 1 in 5)

Gates, Road tunnel

Ferry P　Ferry V

Ferry (passenger), Ferry (vehicle)

RAILWAYS

Track multiple or single

Bridges, footbridge

Track under construction

Level crossing

Siding

Viaduct, embankment

Tunnel, cuttings

Station, (a) principal

Light rapid transit system,
narrow gauge or tramway

Light rapid transit system
station

WATER FEATURES

Marsh or salting

Slopes　Cliff
Towpath　Lock
Aqueduct　Canal
Weir
Footbridge　Bridge
Lake
Canal (dry)

Ford
Beacon
Flat rock
Sand
Normal tidal limit　Dunes

Shingle
Lighthouse
(in use)
Lighthouse
(disused)
Low water mark
Mud
High water mark

Lighthouse

HEIGHTS

1 metre = 3·2808 feet

Contours are at 10 metres
vertical interval

-144

Heights are to the nearest
metre above mean sea level

Where two heights are shown the first
height is to the base of the triangulation
pillar and the second (in brackets) to the
highest natural point of the hill

ROCK FEATURES

Outcrop

Cliff

Scree

PUBLIC RIGHTS OF WAY

Footpath

Bridleway

Restricted byway

Byway open to all traffic

The symbols show the defined route so far
as the scale of mapping will allow.

The representation on this map of any other
road, track or path is no evidence of the
existence of a right of way. Not shown on
maps of Scotland

Danger Area

Firing and Test Ranges
in the area. Danger!
Observe warning notices.

OTHER PUBLIC ACCESS

Other route with public access
(not normally shown in urban
areas). Alignments are based on
the best information available.
These routes are not shown on
maps of Scotland.

On-road cycle route

Traffic-free cycle route

4　National Cycle Network number

8　Regional Cycle Network number

National Trail, European Long
Distance Path, Long Distance
Route, selected Recreational Routes

BOUNDARIES

National

District

County, Unitary Authority,
Metropolitan District
or London Borough

National Park

ANTIQUITIES

+　Site of antiquity

Battlefield (with date)

Visible earthwork

VILLA　Roman

Castle　Non-Roman

TOURIST INFORMATION

Camp site / caravan site

Garden

Golf course or links

Information centre (all year / seasonal)

Nature reserve

Parking, Park and ride (all year / seasonal)

Picnic site

Recreation / leisure / sports centre

Selected places of tourist interest

Telephone, public / roadside assistance

Viewpoint

Visitor centre

Walks / Trails

World Heritage site or area

Youth hostel

LAND FEATURES

Electricity transmission line
(pylons shown at standard spacing)

Pipe line
(arrow indicates direction of flow)

ruin　Buildings

Important building (selected)

Bus or coach station

Current or { with tower
former place
of worship { with spire, minaret or dome

Place of worship

Glass structure

Heliport

Triangulation pillar

Mast

Wind pump, wind turbine

Windmill with or without sails

Graticule intersection at 5' intervals

Cutting, embankment

Landfill site or slag/spoil heap

Coniferous wood

Non-coniferous wood

Mixed wood

Orchard

Park or ornamental ground

Forestry Commission land

National Trust (always open / limited access,
observe local signs)

National Trust for Scotland (always open /
limited access, observe local signs)

ABBREVIATIONS

Br	Bridge	MS	Milestone
Cemy	Cemetery	Mus	Museum
CG	Cattle grid	P	Post office
CH	Clubhouse	PC	Public convenience (in rural areas)
Fm	Farm	PH	Public house
Ho	House	Sch	School
MP	Milepost	TH	Town Hall, Guildhall or equivalent

Magnetic North　Grid North　True North

Diagrammatic only

Scale 1: 50 000

2 centimetres to 1 kilometre (one grid square)

2　1　0 Kilometres　1　2　3

1　0 Miles　1　2

1 kilometre = 0·6214 mile

1 mile = 1·6093 kilometres

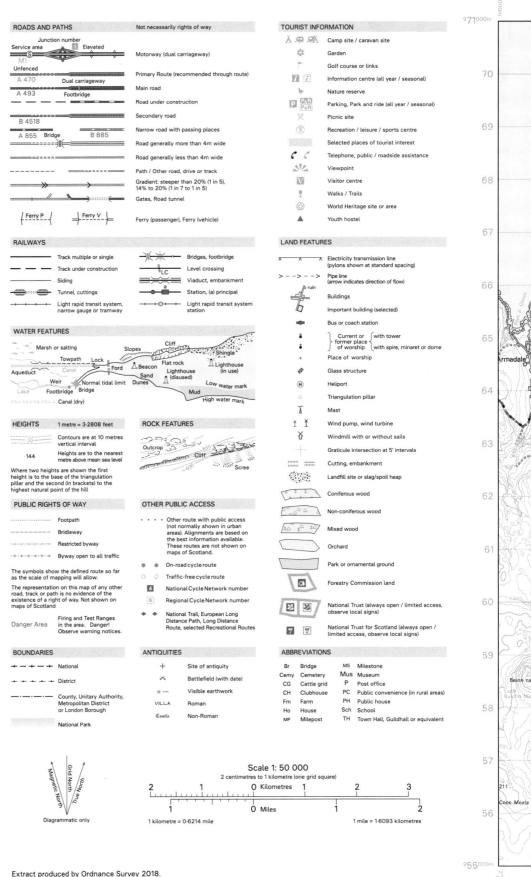

Four colours should appear above; if not then please return to the invigilator.

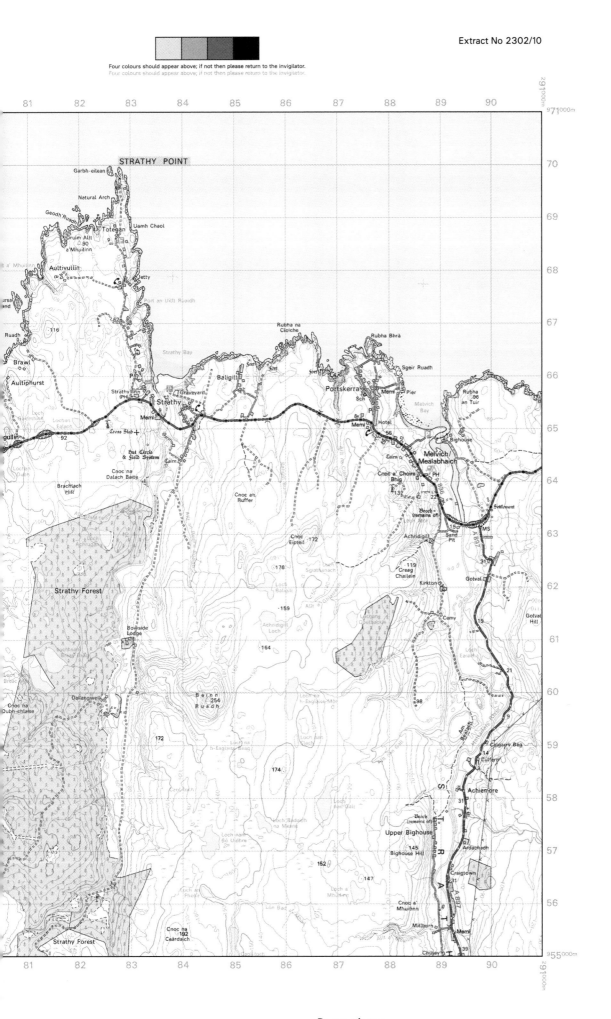

[BLANK PAGE]

DO NOT WRITE ON THIS PAGE

National Qualifications 2018

X833/75/31

Geography
Ordnance Survey Map
Item B

TUESDAY, 1 MAY

1:00 PM – 3:20 PM

The colours used in the printing of these map extracts are indicated in the four little boxes at the top of the map extract. Each box should contain a colour; if any does not, the map is incomplete and should be returned to the Invigilator.

Ordnance
Survey

1:50 000 Scale
Landranger Series

ROADS AND PATHS

Not necessarily rights of way

Junction number
Service area — Elevated
M1
Unfenced
A 470 Dual carriageway
A 493 Footbridge
B 4518
A 855 Bridge B 885

Motorway (dual carriageway)
Primary Route (recommended through route)
Main road
Road under construction
Secondary road
Narrow road with passing places
Road generally more than 4m wide
Road generally less than 4m wide
Path / Other road, drive or track
Gradient: steeper than 20% (1 in 5),
14% to 20% (1 in 7 to 1 in 5)
Gates, Road tunnel

Ferry P Ferry V

Ferry (passenger), Ferry (vehicle)

RAILWAYS

Track multiple or single
Track under construction
Siding
Tunnel, cuttings
Light rapid transit system, narrow gauge or tramway

Bridges, footbridge
Level crossing
LC
Viaduct, embankment
a
Station, (a) principal
Light rapid transit system station

WATER FEATURES

Marsh or salting
Towpath Lock
Aqueduct Canal Ford
Weir
Lake Footbridge Bridge
Canal (dry)

Cliff
Slopes
Shingle
Flat rock
Beacon Sand Lighthouse
Dunes Lighthouse (in use)
(disused) Low water mark
Mud
Normal tidal limit
High water mark

HEIGHTS

1 metre = 3·2808 feet

50 Contours are at 10 metres vertical interval

·144 Heights are to the nearest metre above mean sea level

Where two heights are shown the first height is to the base of the triangulation pillar and the second (in brackets) to the highest natural point of the hill

ROCK FEATURES

Outcrop Cliff
Scree

PUBLIC RIGHTS OF WAY

Footpath
Bridleway
Restricted byway
Byway open to all traffic

The symbols show the defined route so far as the scale of mapping will allow.

The representation on this map of any other road, track or path is no evidence of the existence of a right of way. Not shown on maps of Scotland

Danger Area Firing and Test Ranges in the area. Danger! Observe warning notices.

OTHER PUBLIC ACCESS

• • • • Other route with public access (not normally shown in urban areas). Alignments are based on the best information available. These routes are not shown on maps of Scotland.

● ● On-road cycle route
○ ○ Traffic-free cycle route
4 National Cycle Network number
8 Regional Cycle Network number
◆ ◆ National Trail, European Long Distance Path, Long Distance Route, selected Recreational Routes

BOUNDARIES

National
District
County, Unitary Authority, Metropolitan District or London Borough
National Park

ANTIQUITIES

+ Site of antiquity
× Battlefield (with date)
☆ ···· Visible earthwork
VILLA Roman
Castle Non-Roman

TOURIST INFORMATION

Camp site / caravan site
Garden
Golf course or links
Information centre (all year / seasonal)
Nature reserve
Parking, Park and ride (all year / seasonal)
Picnic site
Recreation / leisure / sports centre
Selected places of tourist interest
Telephone, public / roadside assistance
Viewpoint
Visitor centre
Walks / Trails
World Heritage site or area
Youth hostel

LAND FEATURES

Electricity transmission line (pylons shown at standard spacing)
Pipe line (arrow indicates direction of flow)
Buildings
Important building (selected)
Bus or coach station
Current or former place of worship { with tower / with spire, minaret or dome }
Place of worship
Glass structure
Heliport
Triangulation pillar
Mast
Wind pump, wind turbine
Windmill with or without sails
Graticule intersection at 5' intervals
Cutting, embankment
Landfill site or slag/spoil heap
Coniferous wood
Non-coniferous wood
Mixed wood
Orchard
Park or ornamental ground
Forestry Commission land
National Trust (always open / limited access, observe local signs)
National Trust for Scotland (always open / limited access, observe local signs)

ABBREVIATIONS

Br	Bridge	MS	Milestone
Cemy	Cemetery	Mus	Museum
CG	Cattle grid	P	Post office
CH	Clubhouse	PC	Public convenience (in rural areas)
Fm	Farm	PH	Public house
Ho	House	Sch	School
MP	Milepost	TH	Town Hall, Guildhall or equivalent

Extract No 2303/164

Four colours should appear above; if not then please return to the invigilator.

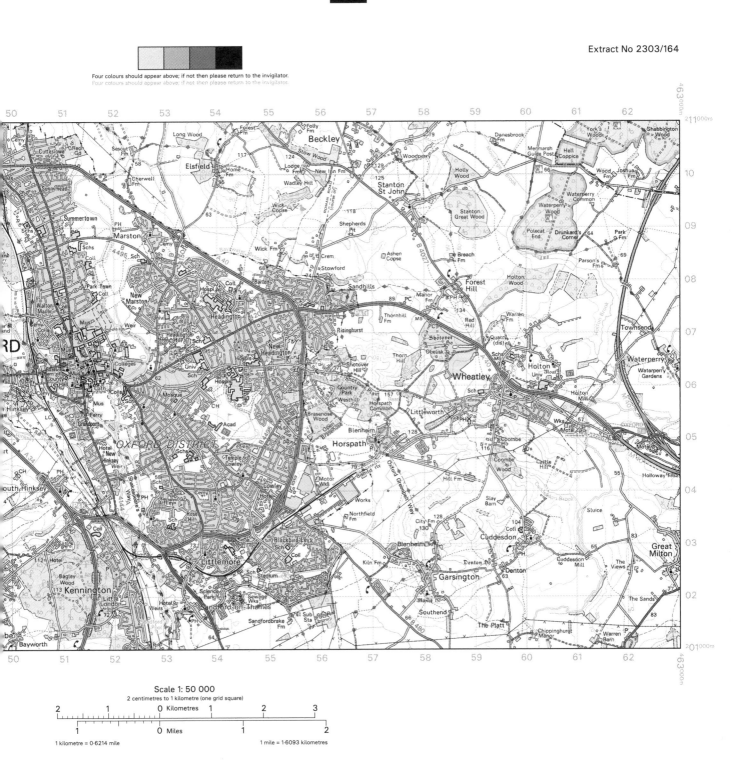

Scale 1: 50 000

2 centimetres to 1 kilometre (one grid square)

Kilometres

Miles

1 kilometre = 0·6214 mile

1 mile = 1·6093 kilometres

[BLANK PAGE]

DO NOT WRITE ON THIS PAGE

General Marking Principles for National 5 Geography

Questions that ask candidates to *Describe* . . . (4–6 marks)

Candidates must make a number of relevant, factual points. These should be key points. The points do not need to be in any particular order. Candidates may provide a number of straightforward points or a smaller number of developed points, or a combination of these.

Up to the total mark allocation for this question:

- **One mark** should be given for each accurate relevant point.
- **Further marks** should be given for development and exemplification.

Question: Describe, in detail, the effects of two of the factors shown. (Modern factors affecting farming).

Example:

New technology has led to increased crop yields (*1 mark*), leading to better profits for some farmers (*second mark for development*).

Questions that ask candidates to *Explain* . . . (4–6 marks)

Candidates must make a number of points that make the process/situation plain or clear, for example by showing connections between factors or causal relationships between events or processes. These should be key reasons and may include theoretical ideas. There is no need for any prioritising of these reasons. Candidates may provide a number of straightforward reasons or a smaller number of developed reasons, or a combination of these. The use of the command word "explain" will generally be used when candidates are required to demonstrate knowledge and understanding. However, depending on the context of the question the command words "give reasons" may be substituted.

If candidates produce fully labelled diagrams they may be awarded up to full marks if the diagrams are sufficiently accurate and detailed.

Up to the total mark allocation for this question:

- **One mark** should be given for each accurate relevant point.
- **Further marks** should be given for developed explanations.

Question: Explain the formation of a U-shaped valley.

Example:
A glacier moves down a main valley which it erodes (*1 mark*) by plucking, where the ice freezes on to fragments of rock and pulls them away (*second mark for development*).

Questions that ask candidates to *Give reasons* . . . (4–6 marks)

Candidates must make a number of points that make the process/situation plain or clear, for example by showing connections between factors or causal relationships between events or processes. These should be key reasons and may include theoretical ideas. There is no need for any prioritising of these reasons. Candidates may provide a number of straightforward reasons or a smaller number of developed reasons, or a combination of these. The command words "give reasons" will generally be used when candidates are required to use information from sources. However, depending on the context of the question the command word "explain" may be substituted.

Up to the total mark allocation for this question:

- **One mark** should be given for each accurate relevant point.
- **Further marks** should be given for developed reasons.

Question: Give reasons for the differences in the weather conditions between Belfast and Stockholm.

Example:
In Stockholm it is dry, but in Belfast it is wet because Stockholm is in a ridge of high pressure whereas Belfast is in a depression (*1 mark*). Belfast is close to the warm front and therefore experiencing rain (*second mark for development*).

Questions that ask candidates to *Match* (3–4 marks)

Candidates must match two sets of variables by using their map interpretation skills.

Up to the total mark allocation for this question:

- **One mark** should be given for each correct answer.
- **Question:** Match the letters A to C with the correct features.

Example: A = Forestry (*1 mark*)

Questions that ask candidates to *Give map evidence* (3–4 marks)

Candidates must look for evidence on the map and make clear statements to support their answer.

Up to the total mark allocation for this question:

- **One mark** should be given for each correct statement.

Question: Give map evidence to show that part of Coventry's CBD is located in grid square 3379.

Example: Many roads meet in this square (*1 mark*).

Questions that ask candidates to *Give advantages and/or disadvantages* (4–6 marks)

Candidates must select relevant advantages or disadvantages of a proposed development and show their understanding of their significance to the proposal. Answers may give briefly explained points or a smaller number of points which are developed to warrant further marks.

Up to the total mark allocation for this question:

- **One mark** should be given for each accurate relevant point.
- **Further marks** should be given for developed points.
- Marks should be awarded for accurate map evidence.

Question: Give either advantages or disadvantages of this location for a shopping centre. You must use map evidence to support your answer.

Example: There are roads and motorways close by allowing the easy delivery of goods (*1 mark*) and access for customers (*1 mark for development*), eg the A46, M6 and M69.

NATIONAL 5 GEOGRAPHY 2017

Section 1: Physical Environments

1. (a) A = arête
   B = corrie
   C = U-shaped valley

   (b) Snow compresses to ice and forms a glacier (1). The glacier uses the process of plucking to steepen the sides of the valley (1). Plucking is when ice sticks onto rocks at the sides of the valley and as the glacier moves downhill, it rips the rocks out (1). Abrasion happens when rocks frozen into the base of the glacier grind at the valley floor as the glacier moves (1). The glacier uses the process of plucking and abrasion to widen and deepen the valley (1). The valley is also weathered above and below the glacier by frost shattering (1). Interlocking spurs are cut off by ice, creating truncated spurs (1).

   Or any other valid point.

   (c) **Farming:** Hill sheep farming is common in a glaciated upland area such as the Cairngorms because sheep are hardy and can survive the cold, harsh conditions (1). The low temperatures and lack of sunshine mean the climate is unsuitable for growing crops (1). Crops are also unable to grow as high rainfall leaches nutrients from the soil (1). The slopes are too steep to use farm machinery (1). Flatter areas on valley floors are often marshy which makes them unsuitable for arable farming (1). Some pastoral farming is possible on valley floors as the grass is better quality (1).

   Forestry: Commercial forestry can take place on the lower slopes of U-shaped valleys where weather conditions are less harsh and soil quality is better (1). This is possible as trees are hardy and can grow on quite steep land and relatively thin soils (1). Trees make use of steep land that is unsuitable for farming or building on (1). Trees help to prevent soil erosion on slopes and flooding in valleys as their roots bind soil together and absorb water (2).

   Recreation and Tourism: Tourists are attracted to glaciated upland areas for the natural scenery which includes ancient forests, vast mountains with glacial features, rivers and lochs (1). Ribbon lochs provide opportunities for water sports such as water skiing and canoeing (1). Mountains provide great opportunities for hill walking and rock climbing (1). Snow-filled corries enable winter sports such as skiing and snow boarding (1). Bird watching is also popular in forests (1). Small settlements eg Aviemore provide tourist services such as hotels, eateries, information centres/car parks/equipment hire shops (1).

   Water Storage and Supply: The high rainfall in upland areas supplies lochs with water that can be used to provide drinking water to settlements (1). The hard impermeable rocks provide excellent geological conditions for water storage in reservoirs (1). Steep sided U-shaped valleys provide natural basins for water storage (1).

   Renewable Energy: Hydro-electric power (HEP) is generated by damming hanging valleys to create electricity using the force of waterfalls (1). Wind turbines can also be located on mountains to take advantage of the windy conditions to generate energy (1).

   Or any other valid point.

2. (a) A = stalactite
   B = grike
   C = joint

   (b) Limestone is made from the decayed remains of sea creatures laid in horizontal layers on seabeds (bedding planes) (1). These sedimentary rocks were uplifted (1) and cracks appeared as the rocks dried out (joints) (1). During glaciation, ice scraped away the topsoil and exposed the bare rock underneath (1). The dry, well-jointed (permeable) bare rock surface allows water to seep down into it (1). Acidic rainwater reacts with the limestone and dissolves the rock (carbonation) (1). The dissolved limestone is carried away by running water (solution) (1). Continued chemical weathering widens and deepens cracks to form gaps called **grikes** (1). Rectangular blocks of limestone called **clints** are separated by the grikes (1).

   Or any other valid point.

   (c) **Farming:** Hill sheep farming is common in upland limestone areas such as the Yorkshire Dales because sheep are hardy and can survive the harsh weather conditions and poor quality grazing (1). Some dairy farms are located on the flat land in the valleys where the soil is more fertile to provide better quality grazing and the weather is warmer and drier (2). A lack of surface water, thin soils and bare rock mean that crops cannot be grown (1).

   Industry: Quarrying is often an important industry in upland limestone areas (1). In the Yorkshire Dales, the main rocks quarried are carboniferous limestone, sandstone and gritstone (1). Cement works can also locate in limestone areas for the raw material lime (1).

   Recreation and Tourism: Tourists visit limestone areas to see the distinctive landscape eg limestone pavements, scars and potholes (1). Visitors like to enjoy the experience of traditional idyllic rural villages (1). People visit limestone caves eg White Scar Caves in the Yorkshire Dales to admire the dripstone features (1). Hill walking in the uplands and cycling in the valleys are common activities (1). Abseiling down limestone scars is a popular activity (1). Many other activities such as caving, pot holing, rock climbing and horse riding are also popular in limestone areas (1).

   Renewable Energy: Upland areas are suitable for generating wind power as they are higher up so more exposed to wind (1).

   Or any other valid point.

3. It is –2°C because it is a high pressure/winter anticyclone (1) and there is often a lack of cloud allowing heat to escape, bringing low temperatures (1). The temperature is very cold because it is December in the UK ie winter time (1). Winds will be gentle as isobars are widely spaced (1). Wind direction is westerly as winds blow clockwise around anticyclones in the UK (1). There is little cloud (1 okta) as cold air sinks in an anticyclone (1). The weather is dry as there are no fronts to bring rain/snow (1).

   Or any other valid point.

4. **Advantages:** Warm, dry and sunny weather improves people's mood (1). People can participate in more outdoor activities such as BBQs (1). Outdoor sports can take place eg tennis matches without being rained off (1). School sports days can safely go ahead due to dry conditions (1). Rising sales of summer goods such as sunscreen and ice lollies increase shops' profits (1).

Disadvantages: Hose-pipe bans enforced due to lack of water (1). Drought conditions reduce the yield of farmers' crops (1). People suffer from sunburn and dehydration (1). More people admitted to hospital with heatstroke (1) putting a strain on resources (1). Forest fires break out (1). Thunderstorms are also a disadvantage of anticyclones (1).

Or any other valid point.

Section 2: Human Environments

5. CBD   – 2573
   Old housing – 2671
   New housing – 2568

6. There is flat land to easily build the houses on (1). The land on the rural/urban fringe is cheaper, so low density housing with gardens/garages can be built (1). There is good road access to this area via the A720/A701/B701 (1) which people can use to commute to their work (1). The area is on the edge of the city, so there will be less noise and air pollution (1) and less traffic, so it will be safer for families (1). There are woods nearby, where residents can go for walks to relax (1). There are also other good opportunities for outdoor recreation such as the ski centre and country park at Hillend (1). There is a Park and Ride scheme nearby which gives easy access to the city (1).

Or any other valid point.

7. **PHYSICAL FACTORS**

Relief: People prefer to live on flat, low-lying areas because it is easier to build on (1). Coastal areas allow trade to take place as ports locate by the sea so many people live nearby (1). Few people tend to live in mountainous areas because steep slopes make it difficult for machinery to operate (1). Upland areas are too cold and wet which makes it difficult to grow crops, so few people live there (1). Mountainous areas also have a low population density because they are often isolated which makes them hard to access (1).

Climate: Many people prefer to live in temperate climates where there is enough rainfall to provide water (1). Few people tend to live in areas with extreme climates because areas like the Sahara Desert with very high temperatures and low rainfall make farming difficult (1). Few people live in areas such as Arctic Canada as permafrost makes building houses and roads difficult as the ground is frozen for much of the year (1). Rainforests have a low population density as they are uncomfortable to live in due to the humid climate (1) and diseases like malaria spread easily (1).

Soil: People prefer to live in areas with fertile soils so that crops can be grown to supply food (1). Where there are poor quality soils eg on steep slopes in Northern Scotland, few crops can be grown so fewer people live there (1). Few people live in hot desert areas because soil dries out and turns to dust, making it difficult to grow crops/keep animals (1).

Natural Resources: Many people tend to live in areas where there are minerals and raw materials to extract and sell (1). Natural landscapes with beautiful scenery attract tourists which generate job opportunities (1) in hotels, shops and restaurants, so more people live in those areas (1). Few people tend to live in areas lacking natural resources because there will be little industry and this means fewer employment opportunities (1).

HUMAN FACTORS

Job Opportunities: Jobs in different industries in urban areas encourage people to move to find work (1). Cities such as Rio de Janeiro have a high population density as there are a variety of job opportunities (1).

Transport and Communications: Areas which are more accessible eg Central Lowlands of Scotland tend to have higher population densities (1). Places with good transport links attract people and industries which in turn create employment opportunities, so more people live there (1).

Services: Towns and cities are crowded as people move to cities like Berlin, London and New York for a variety of amenities and services eg education, health care, jobs and entertainments (1).

Government Aid: Industries locate where there is government funding available, as a result, people move into these areas for work (1). Population density in areas like Syria is falling as people are moving away because of prolonged war (1).

Or any other valid point.

8. For example, in Rio de Janeiro: Wooden shacks have been upgraded to permanent dwellings with some services (1). For example, clean piped water has been provided to help reduce the spread of diseases (1). Residents continually upgrade their homes through a process of "self-help" schemes (1) where the local people are provided with materials like bricks (1). Some prefabricated houses have been built in Rocinha by the Brazilian government (1) with basic facilities like toilets and electricity (1). The residents have been given legal rights to the land where their house is built (1). Roads have been built/improved in the favelas allowing services like rubbish collections to take place (1).

In some favelas cable car systems have been constructed to improve transport for residents (1). There have been some schools and health clinics provided for residents (1). Some charities have also donated money to help improve the standard of living of people in shanty towns (1) eg by providing computers in schools (1). Security has been improved by having more police patrols (1) which have helped to reduce drug-related crime (1).

Or any other valid point.

Section 3: Global Issues

9. (a) Overall the temperatures increased between 1996–2016 (1) by 0.6 degrees (1). The biggest increase was between 1997–1998 where it increased by 0.15 degrees (1). Temperatures decreased only five times in 20 years (1) with the largest decrease between 1998–1999 where it dropped by 0.2 degrees (1). It took seven years for temperatures to reach the same level increase as 1998 at just over 0.5 degrees (1). Temperatures have been continually rising since 2011, reaching the highest level in 2016 with an increase of just over 0.4 degrees (1).

Or any other valid point.

(b) In the UK, the government encourages people to make their houses more energy efficient by giving grants for things like loft insulation, which reduce the amount of energy used (1). Turning off lights, electrical appliances and turning down thermostats reduces the amount of fossil fuels used putting less CO_2 into the atmosphere (1). Many countries like the

UK encourage the use of public transport so reducing the damaging emissions from cars (1). In Brazil laws have been passed to reduce the removal of forest through burning and illegal logging so reducing the amount of CO_2 released into the atmosphere (1). The UK government is trying to reduce the use of fossil fuels such as coal, oil and natural gases by introducing targets for renewable energy using green fuels such as HEP, wind power, solar power (1). Many world nations including the UK take part in Climate Change Conferences, for example the Paris Conference December 2015 where nations agreed targets to reduce the causes of global warming (1). An increasing number of cities are introducing policies to reduce car use and therefore greenhouse gas emissions (1) such as the new tram system in Edinburgh (1) and bicycle-friendly infrastructure in Amsterdam (1).

Or any other valid point.

10. (a) The graph for Canada shows a tundra climate while the graph for Indonesia shows a tropical rainforest (equatorial) climate (1). The tropical rainforest climate is much wetter and much warmer than the tundra climate (1). The wettest month in the rainforest has about 310 millimetres of rain whereas the wettest month in the tundra has about 30 millimetres (1). January and March are the wettest months in the rainforest whereas July is the wettest month in the tundra with some months having less than 10 millimetres (2). The highest temperature in the rainforest is about 27°C but in the tundra it is only 4°C (1). The range of temperature in the tundra is 38°C but only 2°C in the rainforest (1).

Or any other valid point.

(b) **Environment:** In Indonesia, for example, multiple fires have destroyed large areas of the tropical rainforest (1), completely destroying animal habitats and the entire ecosystem (1). Orang utans have been particularly badly affected by this as they are already endangered and have been further threatened by the destruction of their habitat (1). Other animals such as tigers have fewer animals to prey on due to the smaller natural areas of forest which are left and often impact on local communities by taking their livestock instead (2). Many of the fires have been caused by small-scale farmers who want to clear areas of trees to plant them with cash crops such as palm oil (1). Pollution has resulted from burning rainforests and has been so bad that cities such as Jakarta and Kuala Lumpur have been covered in a thick smog, partly caused by rainforest burning (1). The fires also add large amounts of carbon dioxide to the atmosphere leading to further global warming and climate change (1). Road building through the forest not only destroys all the vegetation but opens up new areas to exploitation by small-scale farmers or by logging companies so the construction of roads has an especially bad effect on the forest (2).

People: All of these activities impact also on indigenous peoples who have lived in the forest for generations — they lose their ancestral lands, their food sources, culture and way of life (2). Also when they come into contact with outsiders, they may contract illnesses to which they have little or no resistance, resulting in serious illness or worse (1). Sometimes indigenous communities have had their lands forcibly taken from them with outbreaks of violence resulting in many casualties among the forest peoples (2).

Or any other valid point.

11. (a) Tropical storms (also called hurricanes, typhoons and tropical cyclones) form over oceans within 30° north and south of the equator (1) generally where sea temperatures rise over 27°C (1). They are known as hurricanes where they form over the Atlantic Ocean heading westwards towards the Caribbean/the east coast of Central America/Southern USA eg Florida (1). Tropical cyclones form in the Indian Ocean and move towards Bangladesh/Pakistan/India/Indian Ocean islands such as Mauritius/Madagascar (1). Typhoons form in the Pacific Ocean and South China Sea and affect Australia and countries in South East Asia such as the Philippines, China and Japan (1).

Or any other valid point.

(b) **People:** (eg in Typhoon Haiyan which struck the Philippines in November 2013). The subsequent high seas and flooding resulted in over 6,000 people being killed (1). Whole communities and buildings were destroyed by the intense winds of over 196 mph (1). Over a million people were made homeless and suffered from stress due to loss of possessions and housing (1). Roads and railways were destroyed leading to communication problems and making rescue efforts almost impossible (2). Electricity lines were blown down and people were without power supplies for months (1). People were stranded due to flooding which would have been traumatic (1). Fishing boats and other craft may be damaged causing loss of income (1). There were food and water shortages which led to ill-health (1). As a result of extensive flooding, people may catch water-borne diseases which could be fatal (1). There may be looting of homes, factories and other properties causing tension (1). People lost their jobs in factories that had been destroyed (1). Insurance claims are made resulting in the cost of insurance premiums rising in the future (1). Whilst businesses are closed, earnings (and profits) will be lost (1). Crops were damaged which led to lower productivity and loss of earnings from exports (1). Even after three years, thousands of people in cities such as Tacloban are still living in temporary accommodation, reducing their quality of life (1).

Environment: Extensive flooding occurs as a result of the huge amount of rain which falls during a tropical storm (1). Flooding can lead to sewer systems overflowing and spreading disease (1). There will be structural damage to buildings which may have to be pulled down and rebuilt (1). Sensitive ecosystems may be destroyed and plant and animal habitats lost (1). Fish are often killed in storm surges and because of silting (1). Crops and livestock may be damaged or completely destroyed (1). After Typhoon Haiyan, mudslides were common because the soil was saturated (1). They flowed quickly down hillsides, destroying houses and crops and killing people and livestock (1). In many parts of the Philippines, extensive coastal erosion resulted in loss of farmland and whole communities (1).

Or any other valid point.

12. (a) Between 1995 and 2010 the developing countries' share of world trade has increased, while the

developed countries' share has gone down (1). In 1995 developed countries accounted for just under 70% of world trade, but by 2000 this had dropped to around 68% (1). By 2010 developed countries' share had dropped again by around 13% (1).

In 1995 developing countries accounted for just 28% of world trade, but by 2000 it had risen to around 31% and by 2010 it rose again by 9% (1).

Or any other valid point.

(b) Trade is the exchange of goods and services between countries. More than half the world's trade takes place between just eight countries known as the G8 (1). Usually, developed countries export valuable manufactured goods such as electronics and cars and import cheaper primary products such as tea and coffee (1). In developing countries, the opposite is true. This means that developing countries have little purchasing power, making it difficult for them to pay off their debts or escape from poverty (2). The price of primary products fluctuates on the world market. Workers and producers in developing countries lose out when the price drops, but they benefit when it rises (2). This instability makes it difficult to plan improvement, either locally on farms or in wider government (1). Sometimes developed countries impose tariffs and quotas on imports (1). Tariffs are taxes imposed on imports, which makes foreign goods more expensive to the consumer (1). Quotas are limits on the amount of goods imported and usually work in the developed country's favour (1). Poorer countries supply resources such as timber, agriculture, oil and mining products, often at low prices. These products are used in manufacturing industries to make products which are then sold for large profits, often to poorer countries (2). Often poor countries rely on only one or two raw materials such as Ecuador which grows bananas (1). When the price or demand for bananas falls, the country's income can be badly affected (1). This means countries may need to turn to borrowing and increasing their debts (1).

Or any other valid point.

13. (a) Between 2006 and 2014 the number of tourists visiting Scotland from the USA decreased from 475,000 in 2006 to 275,000 in 2010 (1) and then increased to 418,000 in 2014 (1). The number of tourists coming to Scotland from France, USA, Canada, Ireland and Spain decreased (1). There was an increase in tourists visiting Scotland from Germany, Australia, the Netherlands and Scandinavia (1). The amount of tourists from the rest of the world increased from 976,000 in 2006 to 1,106,000 in 2014 (1). The number of people visiting Scotland from Australia increased over the years from 133,000 in 2006 to 158,000 in 2014 (1). Overall, the total number of tourists visiting Scotland decreased from 2,732,000 in 2006 to 2,700,000 in 2014 (1).

Or any other valid point.

(b) Mass tourism has increased due to improvements in road, rail and air travel which enable people to travel more easily (1). Holiday pay means people can afford to take time off work for a break (1). Increased time off work eg bank holidays gives people the opportunity to visit different places (1). Tour operators and travel agents make it easier to go on holiday abroad due to package deals which often include flights, transfers, meals and holiday reps on hand to solve any problems (2). Cheap package holidays and budget airlines such as EasyJet make holidays more affordable to many people (1). TV travel programmes and adverts on social media inspire people to visit foreign locations (1). The demand to explore various places of interest has increased as globalisation has made the world smaller (1). People also want to experience different cultures and new adventures (1).

Or any other valid point.

14. (a) Mostly there has been a global reduction in worldwide mortality rates from malaria (1), although in South America countries such as Surinam and Venezuela have experienced an increase since 2000 (1). In Brazil and Peru, the mortality rates have decreased by 75% since 2000 (1) and in Argentina the disease appears to have been eliminated since 2000 (1). In Africa most countries have experienced a reduction of between 0% and 74% (1). Nigeria and Kenya are both in this category (1). In SW Africa, countries such as Namibia and Botswana have had a reduction of over 75% in malaria mortality (1).

Or any other valid point.

(b) If **malaria** chosen:

Malaria happens when the parasites injected into the bloodstream by mosquitoes migrate to the liver, multiply and break out in a new form to attack the red blood cells (1). This causes the victim to become seriously ill and if not treated can result quickly in death (1). Symptoms usually start after about a week to 10 days and can include fever, shaking, chills, sickness, vomiting and muscle pains (1). Children under 5 are often worst affected because they have built up less resistance than adults (1). Malaria can recur and so people may often experience several bouts of illness (1). This has a very serious economic effect on their families as, if they cannot work, they may lose income (1). As a result, families may not be able to afford to send their children to school, so they lose out on education (1). Their income may be so low that they cannot afford sufficient food and so malnutrition and hunger can also be a problem (1). Crops may be left unharvested in the fields because farm workers are too ill to gather them in (1). The whole economy of a malaria-affected country can suffer because of low productivity, as much of the workforce is frequently off sick (1). Few tourists want to visit the country because of the threat from malaria, further hitting the country economically (1).

Or any other valid point.

If **cholera** chosen:

Cholera can cause extreme sickness, vomiting, muscle cramps and diarrhoea within 2 to 5 days of infection (1). This can cause the victim to become dehydrated very quickly due to loss of body fluids (1). This can lead to shock, a severe drop in blood pressure and death if not treated quickly (1). Cholera often has the worst impact on areas where lots of people are living close together in insanitary conditions because the bacteria can spread so quickly from person to person (1). This is often the result of a natural disaster such as an earthquake or hurricane or due to war damage (1). Children are at particular risk and can die from cholera within 24 to 48 hours if they don't receive the right treatment (1). Up to

60% of people who develop cholera will die if they are not treated (1). The impact on communities is therefore very high as workers are off sick and productivity is consequently very low (1). This affects the whole economy of the country as resources are used up fighting the cholera outbreak instead of being invested in other areas such as education (1). People who recover from cholera are often weak and have lowered resistance to fight off other diseases, so their long term health suffers (1).

Or any other valid point.

NATIONAL 5 GEOGRAPHY
2017 SPECIMEN QUESTION PAPER

1. (a) Corrie with lochan in squares 7921 & 8021 (1). Scree slopes on south side of corrie (1). U-shaped valley at 8414 (1). Pyramidal peak (Pen-y-Fan) at 012215 (1). Arete at 016213 (1). Corrie with lochan (Llyn Cwn Llwch) at 0022 (1).

      Or any other valid point.

   (b) Glacier forms in corrie/north facing slope and moves downhill due to gravity (1). Eroding sides and bottom of valley (1) through plucking and abrasion (1). Action makes valley sides steeper and valley deeper (1). When glacier retreats a deep, steep, flat-floored U-shaped valley left behind (1). Original river in valley now seems too small for wider valley and is known as misfit stream (1).

      Or any other valid point.

2. (a) From point 905175 river is flowing south (1) down steep-sided V-shaped valley (squares 9017 & 9016) (1). River is joined by tributaries from west such as at 906166 (1). Confluence at 911153 (1) and from this point river gets wider (1). Meander at 911152 (1). At least 3 waterfalls marked in square 9009 (1). In this square river is flowing south-west (1).

      Or any other valid point.

   (b) At river meander, water pushed towards outside of bend causing erosion (1), by processes such as corrosion or hydraulic action (1). Slower flow of water on inside bend causes deposition (1). Over time erosion narrows neck of meander (1). In time, usually during a flood, river will cut right through neck (1). Fastest current is now in centre of river and deposition occurs next to banks (1) eventually blocking off meander to leave ox-bow lake (1).

      Or any other valid point.

3. Answers will vary depending upon the land uses chosen.

   Tourism and recreation examples: Tourists able to visit show caves at Dan-yr-Ogof in 8316 (1) other attractions nearby which might be of interest such as Shire Horse Centre and public house in 8416 (1). Camp site in this square where would be able to stay (1). Lots of opportunities for outdoor enthusiasts such as walking Brecon Beacons Way (7922) (1) nature reserves such as in 8615 where visitors may see rare wildlife species (1). Climbers could tackle the cliffs in Pen-y-Fan (0121).

   Farming: Lots of mountainous land suitable for hill sheep farming (1) sheep able to survive in the colder, windy and wet conditions (1). Farms such as Coed Cae Ddu in 9510 have good road connections, only about 2 kilometres from an A class road, giving good access to markets (1) patches

of woodland which provide shelter for the ewes, especially at lambing time (1). Other farms such as Pwllcoediog (8416) able to benefit from high numbers of visitors by earning extra income from bed and breakfast (1).

Industry: Limestone areas such as Brecon Beacons are sometimes used for extraction of limestone (1). Quarries could be built here as there is limestone and an A class road (A4067) nearby for material to be transported (1). Opencast working shows evidence of industry, grid reference 8211 (1). Works located at 847107 where land is flat so easy to build on (1). Good communication with two main roads A4109 and A4221 close by, also rail link into works (1). Small settlements close by like Dyffryn Cellwen where workers could be found (1).

4. (a) Cape Wrath has north wind whereas Banbury has west wind (1). 35 knots at Cape Wrath but calmer in Banbury at 15 knots (1). Dry in Banbury but snow showers at Cape Wrath (1). 6 oktas cloud cover at Cape Wrath but only 2 oktas over Banbury (1). Temperature at Cape Wrath is much colder at 2°C, while at Banbury is 11°C (1).

   (b) Should not go walking in hills as cold front about to arrive in area (1) which will bring heavy rain showers (1). Will also cause temperature to drop close to freezing point and could be snow (2). If not properly equipped could suffer from the cold and get hypothermia (1) especially as isobars are close together resulting in high wind chill (1). If heavy snow or low cloud they could lose their way easily and need to be rescued (1) these conditions are life-threatening and they should wait for a better day (1).

      Or any other valid point.

5. Tropical continental air mass will bring hot, dry weather in summer which could result in droughts (1). Might need to be hosepipe bans (1). Grass might wither and die causing problems for livestock farmers (1). Ice cream sales might rise (1) as people make most of sunny weather and head for beach (1). Could be very hot and difficult to do physical work outside (1). Heavy rain from thunderstorms might cause flash floods (1).

   Or any other valid point.

6. Answers will vary depending upon the land uses chosen.

   Examples of problems between tourists and farmers: In Cairngorms, tourists can disrupt farming activities as walkers leave gates open, allowing animals to escape (1). Tourists' dogs can worry sheep if let off the lead (1). Stone walls are damaged by people climbing over instead of using gates/stiles (1). Solved by putting in kissing gates or stiles so that people don't have to open gates (1). Noisy tourists can disturb sheep especially during breeding season (1). Farmers may restrict walkers' access at certain times eg lambing season (1). Farm vehicles can slow up tourist traffic on roads (1) parked cars on narrow country roads can restrict movement of large farm vehicles (1). Some of these problems can be resolved by educating public through methods such as publicising Country Code (1).

   Examples of problems between industry and tourists: Tourists want to see beautiful and unusual scenery of Yorkshire Dales but quarries spoil natural beauty of landscape (1). Lorries used to remove stone endanger wildlife and put visitors off returning to area (1). This threatens local tourist-related jobs eg in local restaurants (1). Large lorries needed to remove quarried stone cause air pollution and dust which spoils atmosphere for tourists (1). Quarry companies have covered vehicles with tarpaulins to try and reduce amount of dust (1). Lorries cause traffic congestion on narrow country roads which slows traffic and delays drivers (1). Some quarries have reduced number of lorries by sending limestone by rail instead (1). Peace and quiet for visitors is disturbed by blasting of rock (1). Quarry companies limit frequency and times of blasting to try to reduce impact on local communities (1). Some wildlife habitats may also be disturbed by removal of rock (1).

Or any other valid point.

7. (a) Main roads lead into this square (1) there is a bus station (1) and two railway stations (1) tourist information centre (1) several churches (1) museum (1).

    Or any other valid point.

  (b) Land is flat so easy to build on (1) space available for expansion (1) eg expansion of motor works at 163823 (1). Good transports links like M42 allowing people and products access to and from area (1). Rail link with Birmingham International rail station gives easy access to airport (1). Many road junctions and intersections connecting area to other areas and less traffic congestion as is away from Birmingham city centre (2). Land is on edge of Birmingham so will be cheaper encouraging housing estates like Sheldon to be built (1). Cheaper land allows houses to be bigger with cul-de-sacs, gardens etc. (1) houses can provide a source of labour for the airport, motor works and business park (1).

    Or any other valid point.

  (c) **Area Y:** Some quieter roads such as cul-de-sacs which would be peaceful area to live in (1). Balsall Heath very close to cricket ground (0684) which would be good for recreation (1) and Cannon Hill Country Park (0683) also close by which would give family easy access to pleasant place to walk (1). Children might also enjoy seeing wildlife at Nature Centre in park (1). Area Y is close to centre of Birmingham which might make it easier for parents to get to work if their jobs are there (1) will also be convenient for shopping as lots of variety in CBD (1). Appears to be industrial areas in Area Y (8410) which might provide job opportunities for parents, conveniently close to where they will live (1).

    Area Z: Further out of city, land prices will be cheaper and may be able to afford better house (1). Evidence of lots of modern housing estates with curvilinear road patterns which will be nice environment to live in as will be more garden space (1) and less traffic, making it safer for families (1). Two schools in Area Z which mean children will not have long journey to get there (1). Two golf courses in Area Z, providing recreation opportunities for Russell family (1) and able to get outdoor exercise easily by walking along North Worcestershire Path which passes through Area Z (1). Good transport links into Birmingham for shopping or jobs via main A435 road which leads into CBD and also nearby Park & Ride (1077) next to station where they could travel by train into centre (1). Lots of open space and areas of woodland make the environment/air quality better than many other parts of city (1).

    Or any other valid point.

8. Pesticides reduce disease producing better crops (1) and surplus to trade (1). Fertilisers increase crop yields (1) leading to better profits for some farmers (1) which can lead to increase in standard of living (1). Mechanisation means less strenuous work for farmer (1) and is quicker and more efficient (1). GM crops produce greater yield and are disease resistant so make a greater profit for farmer (1) can reduce cost to farmer of applying pesticides (1) and reduce risk to his health (1). Growing demand for biofuels means higher crop prices and can result in farmer getting higher income (1) and create employment (1).

Or any other valid point.

9. Stage 2 birth rate was still very high, whereas death rate fell quite quickly — caused rapid rise in total population as people were living longer (1). Death rates fell because clean water supplies introduced, reducing spread of disease (1) and at same time proper sewage systems being built which meant water supplies no longer contaminated, reducing number of people falling ill and dying (2). Advances in medicine such as introduction of penicillin helped keep death rates low, as people could be treated for and cured from illnesses which might have killed them in past (2). Birth rates were still high as people were used to idea many children may not survive until adulthood (1) and also children were required to go out and work because of poverty (1). Stage 3 birth rates started to drop much faster as people realised infant mortality was falling and no longer needed to have extra children as insurance policy (2). Standard of living had improved and wasn't necessary to have lots of children to earn income for family any more (1). Also education about family planning was more common (1) and availability and variety of different methods of contraception was better (1). Falling birth rate meant that rate of population increase started to slow down (1). Stage 4 birth rate fallen so low that in some countries is below death rate and so overall population is falling (1). Japan is example of this (1).

Or any other valid point.

10. **Life expectancy:** Very useful development indicator as shows that people in developed countries such as Finland live much longer than in developing countries such as Chad (1). Likely to be because standard of living in Finland is much better (1) and will be much better hospitals, more doctors and more money to spend on medicine (1) as Finland is wealthy developed country which can afford to pay for all of this (1). In Chad, people will have very hard physically demanding lives which may lead to shorter life expectancy (1). They may also live shorter lives on average because of poor nutrition, food shortages and famine (1).

Percentage of workforce employed in agriculture: Can tell you a lot about a country because if a very high proportion of workforce in agriculture shows less developed country (1). Whereas in developing countries such as Mali, 80% of workers are employed in farming because mostly subsistence farmers who have to grow own food (1). Few other places that can get food from or simply can't afford to buy it (2). Also few other industries for people to get jobs in as country is less developed and there is lack of money to invest in setting up new businesses (2). Developed countries such as the Netherlands have very efficient farming industries which require very few workers (1); their economy is highly developed meaning most people are employed in many other jobs and industries which are available and which provide higher incomes than farming (2).

Or any other valid point.

11. (a) Overall trend is that amount of Arctic Sea ice has decreased between 1979 and 2013 (1) from (around) 7 million square km to (about) 5 million square km (1). Fluctuation in extent of sea ice in certain years (1) eg amount of sea ice increased from 3.75 million square km in 2012 to 5 million square km in 2013 (1). Between 2006 and 2007 was a sharp decrease (1) from 6 million square km to 4.25 million square km (1).

Or any other valid point.

(b) Increased temperatures causing ice caps to melt so Polar habitats beginning to disappear (1). Melting ice causes sea levels to rise (1) threatening coastal settlements (1). Increase in sea temperatures causes water to expand, compounding problem of flooding (1). Global warming could also affect weather patterns, leading to more droughts (1) crop failures and problems with food supply (1). Flooding, causing the extinction of species (1) and more extreme weather, eg tropical storms (1). Tourism problems will increase as will be less snow in some mountain resorts (1). Global warming could threaten development of developing countries as restrictions on fossil fuel use may be imposed to slow rate of increasing CO_2 levels (1). In UK, tropical diseases like malaria may spread as temperatures rise (1). Plant growth will be affected and some species will thrive in previously unsuitable areas (1). Higher temperatures may cause water shortages (1).

Or any other valid point.

12. (a) High rates of deforestation occur in Brazil, DR Congo and Indonesia (1). High rates are also prevalent in areas such as Mexico and most of South America (1). High levels of loss more common in developing countries (1). Moderate levels common throughout Europe, northern Africa and Canada (1). Low rates common throughout USA, China, India and Australia (1).

Or any other valid point.

(b) **Management strategies include:** Habitat Conservation Programmes sometimes established in tundra environments to protect unique home for tundra wildlife (1). In Canada and Russia, many tundra areas protected through national Biodiversity Action Plan (BAP) (1). BAP is internationally recognised programme designed to protect and restore threatened species and habitats (1). Reducing global warming is crucial to protecting tundra environment because heating up of Arctic areas is threatening existence of environment (1). Most governments have promised to reduce greenhouse gases by signing up to Kyoto Protocol (1). Many countries have invested heavily in alternative sources of energy such as wind, wave and solar power. These sources of energy are renewable and more environmentally friendly than burning fossil fuels, which increase carbon emissions and global warming (2). Some oil companies now schedule construction projects for the winter season to reduce environmental impact (1). Projects work from ice roads, which are built after ground is frozen and snow covered. This limits damage to sensitive tundra (1). Some oil companies

locate polar bear dens using infrared scanners and do not work within 1.6 kilometres of these dens (1). Number of Arctic research programmes, such as International Association of Oil & Gas Producers' joint industry programme on Arctic oil spill response technology (1). This programme attempts to increase effectiveness of dispersants in Arctic waters, oil spill modeling in ice and use of remote sensors above and under water (2). Many companies operate sophisticated systems to detect leaks (1). Many companies work with local communities to understand and manage potential local impacts of their work (1). Many countries have set up national parks such as Arctic National Wildlife refuge in Alaska to protect endangered animals in tundra (1). Trans-Alaskan pipeline is raised up on stilts to allow Caribou to migrate underneath (1).

Or any other valid point.

13. (a) Over last 100 years, number of eruptions has increased from 43 in 1910s to 70 in 2010s (1). Apart from decades of 1920s, 1970s and 1990s, amount of volcanic activity in each decade increased (1). Least number of eruptions was in 1920s with only 31 (1). Big drop between 1910s and 1920s with a drop of 12 eruptions (1). Also in 1990s there were 12 fewer eruptions than 1980s (1). Biggest increase between 1990s and 2000s with 13 more eruptions (1). Decades with greatest number of eruptions were 1980s, 2000s and 2010s at 66, 67 and 70 (1).

Or any other valid point.

(b) **For Pico de Fogo volcano:** Heat from lava flows set fire to main settlements destroying two villages as well as forest reserve (2) endangering the vegetation and animal habitat (1). Around 1,500 people forced to abandon homes before lava flow reached villages of Portela and Bangeira on Fogo island (1). More than 1,000 people evacuated from Cha das Caldeiras region at foot of volcano to ensure safety and prevent injuries (1). Airport was closed, as ash filled sky, to prevent risk of planes crashing (1). Buildings and records were destroyed resulting in some of history of area being lost (1). Roads and transport routes destroyed, affecting tourist industry on island (1). Volcano destroyed agricultural land which resulted in loss of fertile land (1) decreasing ability of area to produce crops (1) and support local population (1). Tourism might increase as volcano becomes tourist attraction improving economy of the island (1).

Or any other valid point.

14. (a) Europe dominated world trade exports with around 43% in 2005 (1). Dropped to around 38% in 2010 (1). Europe still largest exporter in 2010 (1). Asia had second largest regional share of world trade with around 27% in 2005 (1), growing to around 31% in 2010 (1). Africa's share is low, around 3%, (1) but has grown by about 1% (1). North America's share has dropped from just under 15% in 2005 to around 14% in 2010 (1).

Or any other valid point.

(b) Farmers paid fair wage for their work (1) and safer working conditions promoted (1). Money from fair trade can be used to improve services in local communities (1) such as schools and clinics (1) which improves standard of living (1). More money goes directly to farmer, cuts out middlemen who take some of profits for themselves (1). Farmers receive guaranteed minimum price so are not affected as much by price fluctuations (1). Fair trade encourages farmers to treat workers well and to look after environment (1). Often fair trade farmers are also organic farmers who do not use chemicals on crops so protect environment (1). Health care services and education programmes available and tackle problems of HIV/AIDS(1).

Or any other valid point.

15. (a) USA has six out of ten most popular tourist attractions in world including Niagara Falls and Disneyland (1). Most visited tourist attraction is Times Square in USA with 35 million visitors per year (1). Washington D.C. is second most popular tourist destination with 25 million visitors (1). Trafalgar Square is most popular tourist area in Europe (1). Notre Dame and Disneyland in Paris are most visited attractions in France with 12 million and 10.6 million visitors a year (2). Disneyland Tokyo is most visited attraction in Asia (1). Four out of top ten most popular tourist destinations are Disneyland/Disneyworld parks located on 3 different continents (1).

Or any other valid point.

(b) **People (positive):** Local people employed to build tourist facilities eg hotels (1) and work in restaurants and souvenir shops (1). Employment opportunities allow locals to learn new skills (1) eg obtain foreign language (1) and earn money to improve standard of living (1). Services improved and locals can benefit by using tourist facilities such as restaurants and water parks (1). Better employment opportunities increase the local governments' revenue as wages are taxed (1) so can invest in schools, healthcare and other social services (1). Locals can experience foreign languages and different cultures (1) and can benefit from improvements in infrastructure eg roads and airports (1).

People (negative): Tourist-related jobs are usually seasonal therefore some people may not have income for several months (1) eg at beach and ski resorts (1). Large numbers of tourists can increase noise pollution and upset peace and quiet (1). Local people may not be able to afford tourist facilities as visitor prices are often higher than local rates (1). Tourists can conflict with local people due to different cultures and beliefs (1). There is additional sewage from visitors which increases risk of diseases like typhoid and hepatitis (2).

Environment (positive): Appearance of some areas can be improved by modern tourist facilities (1). Some tourists are environmentally conscious and can have positive impact on landscape by donating money to local projects which help protect local wildlife (1) eg nature reserves (1). Tourist beaches cleaned up to ensure safe for people to use (1) through initiatives like Blue Flag (1). Seas become less polluted as more sewage treatment plants built to improve water quality (1).

Environment (negative): Land lost from traditional uses such as farming and replaced by tourist developments (1). Traditional landscapes/villages spoiled by large tourist complexes (1). Air travel increases carbon dioxide emissions and contributes to global warming (1). Traffic congestion on local roads increases air and noise pollution (1). Tourist facilities such as large high-rise hotels and waterparks spoil look of natural environment (1). Litter causes visual pollution (1). Increased sewage from tourists can cause water pollution (1). Polluted water damages aquatic life and habitats (1).

Or any other valid point.

16. (a) In April 2014 were few cases of Ebola in Africa. By October 2014 were almost 2500 cases in Liberia (1). In Sierra Leone were almost 1,200 cases by October 2014 (1). In Guinea were around 800 cases by October 2014 (1). In Liberia cases rose rapidly from around 250 in August 2014 to around 2500 by October 2014 (1). Sierra Leone witnessed rapid increase in cases from around 500 cases on October 1st 2014 to almost 1200 by mid October 2014 (1).

Or any other valid point.

(b) **Heart Disease:** Lifestyle factors are main cause of heart disease. Many people do not take enough physical exercise which is necessary to keep heart healthy (1). In developed societies many people take car or use lift rather than walking/taking stairs (1). Poor diet leads to heart disease (1). Too much saturated fat can cause hardening or blocking of arteries (1). Many people do not eat enough fruit or vegetables, this can contribute to heart disease (1). Eating too much processed food, with high salt content can also contribute to heart disease (1). Smoking can increase risk of heart disease (1). High stress levels also contribute to heart disease (1). Possible effects of hereditary factors (1).

Cancer: Unhealthy lifestyle is root cause of about third of all cancers (1). Smoking causes almost all lung cancer (1). Poor diet has been linked to bowel cancer, pancreatic cancer and oesophageal cancer (1). Heavy drinking may be a factor in development of cancer (1). Some people may be genetically predisposed to some cancers, eg breast cancer (1). Too much exposure to sun can cause skin cancer (1). Obesity has also been linked with increased cancer risk (1).

Asthma: Infections such as colds or flu affect lungs and narrow airways, making asthma worse (1). Allergic reactions to dust mites in home can cause asthma (1). Pollen from plants outside can cause asthma (1). Traffic fumes in polluted towns and cities can cause asthma (1). Cigarette smoke can cause asthma (1). Asthma can be caused or made worse by damp conditions in home (1). In cases of severe dampness, mould spores may make asthma worse (1).

Or any other valid point.

NATIONAL 5 GEOGRAPHY 2018

1. (a) 827694 – arch   (1)

     812681 – stack   (1)

     843662 – cliff   (1)

(b) Sand spits (long narrow ridges of sand or shingle) form where the coastline changes direction (1). Longshore drift transports sand (1) and deposits it in a sheltered area (1). Deposited sand builds up over time until it is above sea level (1). This deposition continues until the beach extends into the sea to form a spit (1). Sand spits can also develop a hooked or curved end due to a change in prevailing wind/wave direction (1). Mud flats or salt marsh can develop in an area of calm water behind the spit (1).

Or any other valid point.

2. (a) 893618 – ox-bow lake   (1)

     883627 – v-shaped valley   (1)

     895589 – meander   (1)

(b) In its middle and lower course, a river rarely flows in a straight line resulting in water flowing from side to side – meandering (1). The water flows faster on the outside and erodes the outside bend of the river channel to form a river cliff (1). This wearing away of the river banks by the river's load is called corrasion (1). Hydraulic action also takes place where water gets into small cracks forcing pieces to break off the river bed and banks (1). The river flows more slowly on the inside bend and deposits some of its load to form a river beach or slip-off slope (1). Over time continuous erosion of the outer bank and deposition on the inner bank forms a meander in the river (1).

Or any other valid point.

3. A = forestry   (1)

     B = Halladale River   (1)

     C = electricity transmission lines   (1)

4. For Recreation and Tourism:
There are a number of suitable places for people to stay. There are buildings such as Armadale House (790638) (1) which could offer bed & breakfast (1). Tourists would also be able to stay at the caravan & campsite at Melvich (887641) (1). There are a number of nice beaches which tourists would enjoy visiting such as at Strathy Bay (8366) and at Armadale Bay (7964) (1). There would be nice sea views for visitors to enjoy from Strathy Point at 828697 (1) and they might also be able to go fishing from the jetty at 831678 (1).

For Renewable Energy:
This area could be suitable for wind turbines as there are a number of hilly areas where there would be stronger winds (1). It is also a coastal area where winds tend to be stronger (1). There are a number of tracks which would provide access to hilly areas such as south of Bowside Lodge in 8360 (1). As it is a coastal area, it might be suitable for offshore wind power or wave power (1). An exposed area like this in the north of Scotland is likely to have stronger winds and therefore bigger waves (1). The pier at 883657 would provide local access for boats to maintain marine renewable devices (1). Strathy forest has the potential to provide wood for biomass (1).

For Forestry:
The land is not very flat, quite wet in places and so is not good for farming but trees can still be planted here (1) and they clearly grow quite well in this area as there are a number of coniferous plantations already (1) such as at Strathy Forest (8261) (1). Villages such as Melvich and Portskerra might be able to provide a labour force for maintenance and felling (1).

Or any other valid point.

5. Answers will vary depending on the features chosen:

e.g. A number of conflicts have emerged around the Dorset coast.

Strategies include the increase in use of cycle routes and train lines which has helped to reduce traffic congestion on coastal roads (1). Giving the area World Heritage Site status emphasises its worldwide importance helps to protect the coastline (1). Turning a former abandoned quarry into the Townsend Nature Reserve has helped to protect wildlife, flora and fauna (1). It has also been designated a site of special scientific interest (SSSI) which has 7 different species of orchid (1). Marram grass is widely planted to conserve coastal vegetation and reduce the effects of human and physical erosion (1). Lulworth Estate has provided a large car park which has had the effect of reducing some of the parking issues (1). Charges for the car park are spent on improving services in the area which benefit both locals and visitors (1). A bus service has been provided from the nearest train station to encourage visitors to leave the car behind (1). A roundabout has been built at the car park entrance to allow traffic to turn and reduces congestion (1). Lulworth Estate also plans to screen the holiday park to reduce the visual impact on the landscape (1).

Footpath erosion has been resolved be placing limestone cobbles on paths to make them more durable (1). On steep descents wooden steps have been included to prevent further erosion (1). Reseeding and re-routing paths has protected particularly worn areas (1). Lulworth has no bins to encourage tourists to take litter home (1) and the local estate uses funds from the car park to educate and provide guided tours for tourists (1).

The MoD have agreed to avoid using the coast on the busiest days of the year: this reduces the impact on tourist experience although there are times when the coast remains restricted (1). The MoD has provided signage to inform visitors when the coastline is closed and which particular areas of paths cannot be used (1). The firing range reduces its practice during peak times to limit the noise from the range (1). The MoD argue that restrictions on access has preserved the natural beauty of the area (1).

Or any other valid point.

6. Latitude: places in southern England are warmer because they are nearer the Equator (1); temperatures generally decrease the further north you go because the sun's rays are less concentrated further away from the equator (1). It's colder in northern Scotland because there is more atmosphere for the sun's rays to pass through (1)

Altitude/Relief: upland areas are colder as temperatures decrease 1°C for every 100 metres gained in height (1) and wind speeds increase as altitude increases which can affect temperature (1).

Aspect: in the northern hemisphere south facing slopes can be warmer because they face the sun (1). North facing slopes are shaded from the sun and are therefore cooler (1).

Continentality/Distance from the sea: places closer to the sea have warmer/milder winters and cooler summers because oceans heat up slowly in summer and cool slowly in winter (1) oceans act as 'thermal reservoirs' (1) whereas places further inland have a greater annual range in temperature due to distance from the effects of the oceans (1). The North Atlantic Drift keeps the temperatures warmer on the west coast than on the east coast of the UK (1).

Or any other valid point.

7. Due to the approach of the warm front, Stirling's air pressure will fall (1) cloud cover will increase (1) and steady rain will fall (1). Winds will be stronger as the isobars are closer together (1). Because Stirling will be in the warm sector of a depression, temperatures will rise (1) and it will be mild with some cloud cover and occasional showers (1). Due to the cold front arriving, cloud cover will increase (1) with cumulonimbus clouds bringing heavy rainfall to Stirling (1). Temperatures will drop as the cold front passes over (1). As the front begins to move away, the sky will become clear (1) rain will stop (1) air pressure will begin to rise (1).

Or any other valid point.

8. A. From the public telephone in Henwood (4702) to the school near Rose Hill (5303) 6.25 km. (1)

   B. From Forest Farm (5410) to the church in Stanton St John (5709) 3.75 km. (1)

   C. From Waterperry Gardens (6206) to the College (5502) 8.25 km. (1)

9. (a) Area X – 5106 is the CBD as it has old churches (1) main roads leading to the CBD (1) there are museums here (1). Tourist Information centre (1). Bus station (1)

   Area Y – 5502 is a modern suburb as it has a modern street pattern (cul-de-sacs) (1). It is located at the edge of the city as would be expected of the suburbs (1) there are two schools nearby for children of families living in the nearby housing areas (1) there are few main roads, only B class and minor roads (1). There is more open space than would usually be found in either the CBD or inner city (1).

   Or any other valid point.

   (b) **Advantages**
   The land is flat, so easier to build on (1). There is reasonable flat land nearby for expansion or car parking if needed (1). Oxford is nearby, so there is a market for the shopping centre (1). People living in nearby areas, such as New Headington could provide a workforce (1). The A40 is close by to provide easy transport to the area (1). The land is on the outskirts of Oxford so should be cheaper to buy (1). Traffic congestion is also likely to be less of a problem outside of the CBD (1).

   Disadvantages
   There is a river running through Area Z and this could limit the land available for the development or increase building costs (1). Home Farm (GR 541100) may object to the plans (1). There is a small forest (537099) which would cost money to clear or may cause objections to be raised (1).

   Or any other valid point.

10. Answers will vary depending on case study chosen.

    Much of Glasgow's housing stock was run down and in need of repair so the government invested money to renovate the old tenements by putting in new windows, bathrooms and double glazing (1). Some of the poorer housing/tower blocks were pulled down (1) and replaced by new housing to improve living conditions in regenerated areas like Glasgow harbour (1). Old grid iron streets plus increased car ownership caused congestion problems so new transport links like the Partick Interchange was built (1). To try to improve unemployment the government invested in the service sector with many jobs being created in call centres (1). Small industrial units replaced the old heavy industries improving employment in the area (1). Tourism has been encouraged with many new hotels appearing in the gap sites left by the demolished factories (1). Improvements made to the environment by landscaping and improving docklands (1).

    Or any other valid point.

11. In the Lower Ganges Valley, India, new technology such as tractors allows farmers to increase speed and efficiency (1) which provides better profits for some farmers (1). This money can then be used to improve the overall standard of living of the farmers (1). There is less physical work for people (1) but fewer jobs available (1). This can lead to rural depopulation in some areas such as people leave to find work (1). Machines are expensive and not all farmers can afford them, leading to inequality (1).

    The use of irrigation channels can allow two to three harvests a year instead of one, this increases profits for farmers (1). However, as the land is constantly in use the soil quality becomes poorer over time (1).

    The increased use of machinery and chemicals has created new industries and jobs, e.g. mechanics to fix tractors (1).

    The introduction of GM crops can give the farmer a more reliable harvest as the seeds are designed to resist disease (1). Crops can be grown in adverse conditions, e.g. with less water, ensuring a better food supply for the people (1). However, the increased use of fertilizers and pesticides can damage the environment if they get into the water (1). Farmers become reliant on multi-national companies (1).

    Or any other valid point.

12. (a) In 2015 India, China and the USA had the highest Gross National Income (GNI) in the world (1). The GNI of the USA was greater than $4.93 trillion (1). Most African countries for which we have data, had a GNI less than $1.06 trillion (1). The GNI for Brazil was between $2.75.06 – $4.93 trillion (1). Most other South American countries earned less than $1.06 trillion (1). Scandinavian countries all had a GNI of less than $1.06 trillion (1). Germany had a GNI of between $2.75 – $2.93 trillion (1). The UK's GNI was between $1.06 – $2.75 trillion (1).

    Or any other valid point.

    (b) Answers will vary depending on choice.

    If number of people per doctor chosen:
    A high number of people per doctor shows a lack of healthcare provision (1). The more people per doctor the less developed a country will be because there isn't enough money to educate them (1). Developing countries often have a poor education system and lack of universities to train qualified doctors (1). Governments in developing countries cannot afford to keep hospital stocked with adequate provisions (1).

If number of births per 1,000 women per year is chosen:

The lower the number of births per women the more developed a country will be because there is a low infant mortality rate and women do not need to have 'extra' children to ensure some survive (1). Children are not needed to work on the land so birth rates are low (1). Contraception is widely available and family planning clinics allow women to plan for a baby (1). Sex education in schools helps to prevent unwanted pregnancies (1).

If percentage of people working in agriculture is chosen:

The lower the percentage of people working in agriculture the more developed a country will be because most people work in factories or services (1). Developed countries have fewer people working in farming because they can afford to import food from other countries (1). People work in mainly secondary and tertiary industries as there is more money to be made in these sectors (1). More people work in agriculture in the developing world because of the lack of mechanisation (1).

Or any other valid point.

13. (a) In 1990 carbon dioxide accounted for around 24,000 million tonnes of emissions. This had risen to around 34,000 in 2010 (1).

In 1990 methane + nitrous oxide accounted for around 10,000 million tonnes of emissions. This had risen to around 12,000 in 2010 (1).

The change in carbon dioxide emissions is more than methane + nitrous oxide or nitrous oxide with a change of around 10,000, compared to 2,000 for methane + nitrous oxide (1). Greenhouse gasses have increased from 1990–2010 (1) from 34,000 million tonnes to 46,000 million tonnes (1).

Or any other valid point.

(b) **Physical Causes:**

Fluctuations in solar activity over time can increase or decrease global temperatures (1).

The Little Ice Age of 1650–1850 may have been caused by a decrease in solar activity (1).

Volcanic eruptions can impact on global temperatures as large quantities of volcanic dust in the atmosphere shield the Earth from incoming insolation which lowers global temperature (1). For example, the eruption of Mount Pinatubo in 1991 caused a dip in global temperatures when 17 million tonnes of sulphur dioxide were released into the atmosphere (1). This reduced global sunlight by 10% and resulted in a 0.5% temperature decrease globally (1).

Large eruptions may also enhance the greenhouse effect and lead to global warming in some instances (1).

Milankovitch cycles or variations in the tilt and/or orbit of the Earth around the Sun affect global temperature (1). More tilt means warmer summers and colder winters, less tilt has the opposite effect (1).

Changes in oceanic circulation such as the periodic warming (El Nino) and cooling (La Nina) of areas of the tropical Pacific Ocean can impact on global temperature (1).

Melting permafrost from Arctic areas can release large quantities of the greenhouse gas, methane (1).

This exacerbates the natural greenhouse effect, increasing global temperatures (1).

Human Causes:

The burning of fossil fuels produces carbon dioxide which leads to global warming (1).

Car exhausts, nitrogen fertilisers and power stations all produce nitrous oxide which increases the amount of greenhouse gasses in the atmosphere (1).

Worldwide deforestation also increases carbon dioxide levels, by reducing the storage of carbon (1).

CFCs found in fridges, air conditioning and aerosols contribute to global warming (1).

Increases in rice production and cattle farming contribute to atmosphere pollution (1). Because methane is a stronger greenhouse gas, small increases have a larger impact (1).

Or any other valid point.

14. (a) **Deforestation levels:**

Deforestation rose from just under 1.2 billion hectares in 1900 to around 1.8 billion hectares in 2010 (1). Deforestation levels have steadily increased (1). From 1900 to 1980 deforestation increased from around 1.2 billion to 1.6 billion hectares (1) a difference of 0.4 billion (1).

Population:

World population rose from around 0.9 billion in 1900 to just under 7 billion in 2010 (1). Population increased relatively slowly from 1900 to 1950 from around 0.9 billion to 1.1 billion (1).

Or any other valid point.

(b) **Rainforest:**

Plants such as fan palms have large leaves that are good for catching sunshine and water (1). The leaves are segmented, so excess water can easily drain away (1).

Rainforests have a shallow layer of fertile soil, so trees only need shallow roots to reach the nutrients (1). However, shallow roots can't support huge rainforest trees, so many tropical trees have developed huge buttress roots (1). These stretch from the ground to two metres or more up the trunk and help to anchor the tree to the ground (1).

Lianas are woody vines that start at ground level, and use trees to climb up to the canopy where they spread from tree to tree to get as much light as possible (1).

Strangler figs start at the top of a tree and work down. Gradually the fig sends aerial roots down the trunk of the host, until they reach the ground and take root (2). The figs branches will grow taller to catch the sunlight (1) and invasive roots rob the host of nutrients (1). Eventually the host will die and decompose leaving the hollow but sturdy trunk of the strangler fig (1).

Some plants grow thick leaves with drip tips and waxy surfaces to allow water to drain quickly to prevent rotting (1).

Some plants called 'epiphytes' get food from the air and water, and their roots hang in the air, e.g. orchids (1). Trees grow fast and straight to compete for sunlight (1).

Any other valid point.

Tundra:

Successful plants in the tundra are low growing, compact and rounded in order to help protect from the wind (1). Many grow close together for added protection from the weather (1).

The trees that can survive in the tundra are often small (1) and the snow acts as insulation for the trees and helps them stay warmer during the winter months (1).

During winter months, many plants go dormant to tolerate the cold temperatures (1). By going dormant during the winter, plants are able to save energy and use it during more favourable conditions, like the warmer summer months (1).

Plants grow rapidly during the short summer season, and they flower more quickly (1).

The flowers of some plants increase their heat efficiency by slowly moving during the day to position themselves in a direction where they can catch the most rays from the sun (1). Some plants have cup shaped flowers to trap the sun (1). Other plants have protective coverings, such as thick woolly hairs, that help protect them from wind, cold and desiccation (1).

A small leaf structure is another physical adaptation that helps plants survive. Plants lose water through their leaf surface. By producing small leaves the plant is more able to retain the moisture it has stored (2).

Cotton grass has narrow leaves helping to reduce transpiration (1) its dense flower heads reduce heat loss and darker leaves help absorb energy from the Sun (2).

Or any other valid point.

15. (a) There was a general rise, before a spike in 1995, resulting in around $155 million (1). There was a drop to $30 million in 2001 (1), before rising to $220 million in 2005 (1). There was a high of $360 million in 2011, before it dropped to $150 million in 2012 (1). Overall, there was a rise in damage cost from 1990 to 2012 (1).

Or any other valid point.

(b) Answers will vary depending on the case study chosen.

Scientists can monitor seismic activity. Tremors can give warnings as to an imminent eruption (1). If people are warned they can evacuate (1). Scientists successfully predicted the eruption of Mount St Helens in 1980 by measuring the frequency of earthquakes on the mountain. This enabled many locals to escape to safety (1).

Scientists can also monitor gas emissions, such as sulphur dioxide to predict an eruption (1). However, this is an inexact science and scientists can rarely predict too far in advance. Despite scientists noticing increased tremors and sulphur dioxide emissions before the eruption of Mount St Helens, scientists thought that it might still be a few weeks away (1).

Volcanoes sometimes expel lava bombs before an eruption, this would give the population warning to evacuate (1).

Temperatures around the volcano tend to rise as activity increases. Thermal imaging techniques and satellite cameras can be used to detect heat around a volcano (1).

Volcanoes such as Mount St Helens in the USA and Mount Etna in Italy are closely monitored at all times. This is because they have been active in recent years and people who live nearby would benefit from early-warning signs of an eruption (1). Tilt meters which record changes in the shape of a volcano can also give early warning of an eruption (1).

People living in the shadow of a volcano have emergency plans in place and emergency supplies such as bottled water and tinned food are stockpiled to ensure they have vital supplies to survive in the event of an eruption (2).

In the event of a serious eruption, short term aid in the form of food, medicine and shelter could be sent to the area to treat the injured (1). In the case of Mount St Helens a 5-mile exclusion zone was enforced (1).

When the Pico de Fogo Volcano in Cape Verde erupted in 2014, more than 1,000 people were evacuated from the Cha das Caldeiras region at the foot of the volcano immediately after it first erupted (1). Officials closed the airport as the skies darkened with ash to prevent damage to aeroplane engines (1).

Or any other valid point.

16. (a) The number of people employed in the production Fair Trade flowers and plants has decreased by 1,000 (1) from 51,000 to 50,000 (1). Employees in seed cotton have decreased from 60,000 to 53,000 (1). Cocoa producers have increased by 5,000 (1) from 180,000 to 185,000 (1). The number of employees in the growth of Fair Trade tea has increased by about 65,000 people in one year (1) from 300,000 to 365,000 (1). Tea had the largest increase in the number of employees (1).

Or any other valid point.

(b) Answers will vary depending on case study chosen.

Farmers are paid a fair wage for their hard work producing Fair Trade tea in Kenya (1) and safer working conditions are promoted to prevent accidents and injuries (1). Fair trade also encourages farmers to treat their workers well (1). Farmers receive a guaranteed minimum price for their cocoa in Cote d'Ivoire so they are not affected as much by price fluctuations (1) and can receive some money in advance, so they don't run short (1). More money goes directly to the farmer as the 'middle man' is removed (1). Money from Fair Trade bananas in Latin America can be used to improve services in local communities such as schools and clinics (2) which improves peoples' standard of living (1).

Or any other valid point.

17. (a) There are more endangered world heritage sites located in Africa than in any other continent (1). There is 1 world heritage site in danger in the UK (1). Madagascar has a site that is endangered (1). The Middle East has the second highest number of sites including locations in Afghanistan and Iraq (1). There is also one world heritage site in danger in the USA (1). There are 3 sites in danger in South America (1). Central America has 4 world heritage sites in danger (1). Indonesia has 1 site under threat (1).

Or any other valid point.

(b) National parks are set up to protect fragile environments and to encourage sustainable economic development, including tourism (1). Local guides educate visitors on the importance of conservation (1) and can show them projects where their money is being spent to protect the environment (1). Limited numbers of people are allowed access to eco-tourist areas (1) e.g. in Peru daily numbers are restricted on the Inca Trail (1). Tours must be small-scale so companies have to limit group sizes to lessen environmental impact (1). Eco-tourists must follow local customs and respect local cultures (1) e.g. removing shoes before entering temples in Cambodia (1). Tourists are encouraged to follow the code 'take nothing but photographs, leave nothing but footprints' (1).

Or any other valid point.

18. (a) Answers may include:

HIV/AIDS is most prevalent in developing countries (1). The HIV/AIDS rate is over 10% in South Africa, Namibia and Botswana (1). Kenya and Tanzania have a rate of between 6–10% (1). Mauritania, Mali and Ghana have a rate of between 1–5% (1) South and Central American countries like Brazil and Mexico also have low rates of under 1% (1).

Or any other valid point.

(b) Answers may include:

AIDS is a debilitating disease which means that eventually those infected will not be able to work (1). This lowers productivity and hampers development of a country (1). This in turn leads to fewer jobs and less wealth in a country (1).

The death rate will increase and life expectancy decreases (1).

In areas where AIDS is endemic e.g. South Africa or Uganda, children may be left without parents and brought up by grandparents (1), meaning entire middle-aged populations may be missing from societies (1). Those affected will be mainly in the economically active group so the dependency ratio will increase; there will be less people to support the young and elderly (2).

With more adults ill and unable to work then the economically active population reduces (1), resulting in a shortage of labour (1).

Less food will be produced as less people are able to work the land (1).

There may be a loss of tourist revenue if there are known to be specific problems with disease in the area (1).

The young often become carers, therefore missing out on education (1). There will also be a large number of orphans and dissolved families (1).

Relatives of sufferers may be ostracised by their communities (1).

Lack of staff in schools means that many people don't receive enough education about AIDS (1).

Or any other valid point.

Acknowledgements

Permission has been sought from all relevant copyright holders and Hodder Gibson is grateful for the use of the following:

Image © De Visu/Shutterstock.com (2017 page 7);
Image © Fabio Lamanna/Shutterstock.com (2017 page 11);
Image © Richard Roscoe, Photovolcanica (2017 SQP page 15);
Image is reproduced by kind permission of the Fairtrade Foundation © David Macharia (2017 SQP page 16);
Image © jan kranendonk/Shutterstock.com (2017 SQP page 17);
Image © mountainpix/Shutterstock.com (2018 Section 1 page 7);
Image © Martin Kemp/Shutterstock.com (2018 Section 1 page 7);
Image © Christopher Elwell/Shutterstock.com (2018 Section 1 page 7);
Image © chris2766/stock.adobe.com (2018 Section 1 page 7);
Image © cybrain/Shutterstock.com (Elements of this image furnished by NASA) (2018 Section 3 page 17);
Image © Silken Photography/Shutterstock.com (2018 Section 3 page 19);
Image © Hugh Lansdown/Shutterstock.com (2018 Section 3 page 19);
Image © Daimond Shutter/Shutterstock.com (2018 Section 3 page 19);
Image © Bildagentur Zoonar GmbH/Shutterstock.com (2018 Section 3 page 19);
Image © Jack Cronkhite/Shutterstock.com (2018 Section 3 page 19);
Image © nalongsak hoisangwan/Shutterstock.com (2018 Section 3 page 19);
Image © Leonid Ikan/Shutterstock.com (2018 Section 3 page 19);
Image © BMJ/Shutterstock.com (2018 Section 3 page 19);
Ordnance Survey maps © Crown Copyright 2018. Ordnance Survey 100047450.